Praise for *The YouTube Formula*

"Only one or two other people in the entire world understand YouTube on the level that Derral does. I have one of the top three most-watched channels on YouTube, and I still talk about YouTube data and strategy with Derral because nobody is on the same wavelength as he is. Whether you have 5 subscribers or 5 million, you can learn how to be successful on YouTube by reading Derral's book."

—Jimmy Donaldson,
MrBeast

"Study it . . . understand it . . . dream it . . . Derral Eves just wrote the bible to YouTube success."

—Jordi van den Bussche,
Kwebbelkop

Derral has the rare experience of being successful on YouTube as a creator, a marketer, an advertiser, and an entrepreneur. This book teaches you how to understand YouTube from the big picture all the way down to the important details.

—Chad Wild Clay and Vy Qwaint,
Cocreators of Spy Ninjas

"When you talk about the 'brains behind the throne' among top YouTube stars, you're talking about Derral Eves. His unique understanding of the inner workings of YouTube, coupled with an insatiable appetite to find out why certain videos perform and others languish has placed him at the forefront of online video experts. Derral understands not only WHAT you need to do to build a bigger audience, but WHY you need to do it. Whether you're looking to build a big business on YouTube or just build a community of passionate fans, this book packs more inside than you'd get from months of expensive consulting and painstaking trial and error. Find out why so many top YouTubers call Derral both a friend and a confidant by buying this book right now—and you'll be taking the first step to making all your YouTube dreams come true!

—Jim Louderback,
GM VidCon

"Derral Eves has a YouTube super power. He can spend five minutes looking at a YouTube channel's analytics and give a handful of actionable suggestions. Not just helpful tips, but algorithmic game changers. The fingerprints of Derral's strategic brain are on most of my largest successes."

—Jeffrey Harmon,
Cofounder at Harmon Brothers and VidAngel

"Derral is the preeminent expert authority on video marketing in the modern era of digital and social platforms. He has consulted the top YouTube creators and platforms and built his own content channels generating millions of followers. *The YouTube Formula* is a must read for anyone interested in gaining prominence in or simply learning about YouTube content creation, the mysterious algorithm, or optimizing audience growth."

—Ricky Ray Butler,
CEO BEN Group

"Derral has been analyzing the YouTube algorithm longer than anyone I know. Whether you're just starting out or looking to reach ten million subscribers, Derral's book contains specific ideas that can help you do much better than you would on your own."

—Mark Rober,
NASA Engineer turned YouTube Creator

THE
YOUTUBE
FORMULA

DERRAL EVES

THE

YOUTUBE FORMULA

HOW ANYONE CAN UNLOCK

THE ALGORITHM TO DRIVE VIEWS,

BUILD AN AUDIENCE, AND GROW REVENUE

WILEY

Library of Congress Cataloging-in-Publication Data is Available:

ISBN 9781119716020 (Hardcover)
ISBN 9781119716044 (ePDF)
ISBN 9781119716037 (ePub)

Cover Design: Derral Eves
Cover Image: Doodles: Derral Eves
Ripped paper: © stockcam/iStockphoto

SKY10023978_011221

To my wife Carolyn:
No goal or accomplishment would mean anything
without you by my side.
You are my everything . . . for eternity!

CONTENTS

FOREWORD

Everyone should have a YouTube channel. Literally everyone, but especially brands. When I see brands that don't have a presence on YouTube, I think they're insane. It's unfathomable that anyone isn't capitalizing on the opportunity there. It's the most coveted job in America, and with good reason. It is quite literally a gold mine.

When I was a kid, I watched YouTube all the time. It was always my dream job. I didn't want to be an astronaut or a doctor—I couldn't envision a world where I wasn't a YouTuber. I started my channel in 2012 and only got 40 subscribers my first year. Now I have one of the fastest growing channels in the world. I gained more than 15 million subscribers in 2019 alone with just over 4 billion video views. And it's still growing every day.

If you are just getting started on YouTube, do not expect to pull any type of viewership in your first year. If this isn't something you can accept, don't start. But if you can, then you need to do this: make 100 videos. It doesn't matter what they are because they will be terrible, but do something you like doing. Your first 10 videos will be garbage. Then make 10 more. These will also be garbage, and so will the next 10. But eventually, things will start to improve. You'll get better little by little. The best way to improve your content is to make content and see what people like.

Then you'll notice something with your 101st video. It will be in a whole different league from your first video. You will still be a long way from good content, but it will be better than your first video was.

It took me hundreds of videos over several years before I got good at it. I had been making YouTube videos for two years and still didn't make good quality videos, even though I thought they were good at the time. I was like so many creators: I thought the algorithm hated me because I wasn't getting subscribers and views. But in reality, my content wasn't good enough. In fact, my videos were horrible, like most YouTube videos are. Most YouTubers have their priorities backward. They spend all their time thinking about the algorithm in their first hundred videos when they should really be thinking about how to make better content geared toward the viewer.

Unless you're the rare YouTuber who has created content professionally, you're probably average (at best) at making content. You don't go from entertaining nobody to entertaining millions in a day. It's hard work and a slow progressive workup, and it should be. Because if you got millions of viewers overnight, you wouldn't know how to handle them.

Every video is your competitor, so you have to keep putting in the effort to keep your content relevant and competitive. Viewers have to pick between videos every time they go to YouTube, and they'll choose the better video without a second thought. Make your video the better video.

If a video from your channel from six months ago doesn't make you want to barf because of how much better you are at making content now, you're doing it wrong. I can't even stand watching my videos from six months ago because I am so much better at making content now. I can see how bad my old videos are. They could have been so much better! I'm depressed even thinking about those videos.

There is a huge opportunity on YouTube for brands. If every brand knew what I know, they could get a hundred times the viewership they get on television for the same cost. They should be giving creators like me their advertising budget to make YouTube videos around their brand. I'll give you an example. If I was Coca-Cola, I could build a pyramid of a million Coca-Cola cans. Or I could make a video called, "I Filled My Friend's House with a Million Cans of Coca-Cola." These are both banger ideas that would get a ton of views if done right. (This was free, Coca-Cola; next time you can pay me, lol.) Videos like these could lead to millions upon millions more return than TV-spends, and it's exponentially cheaper. And seriously, who really watches TV commercials anyway. Everyone picks up their phones during commercials.

If brands want to get real visibility, they have to get off of TV. What do you think people do—especially anyone under the age of 30—when the ads start rolling on TV? They pull out their phones and quit paying attention! They're not watching your commercial! Their attention has been moved to Twitter. SnapChat. YouTube. Tik-Tok. Tweets spike during TV commercials. And people don't put their phones down until the commercials are over and their show or sporting event comes back on. Three times fewer people skip a brand deal on YouTube than they do on TV, because people are usually watching from their phones. They don't have a second phone to grab when they come across an ad or brand deal on YouTube, so they watch it. And they're watching a YouTuber they watch all the time, so they are also less likely to skip because they trust these YouTubers.

I have more eyeballs on my content every week than the Super Bowl. And I have a personal connection with my viewers who have built trust in me. This audience trust value is worth so much more than an impersonal celebrity or athlete appearance in a TV commercial. Brands shell out more than $5 million for 30 seconds crammed in with a bunch of other ads during the Super Bowl. With YouTubers like me, you could get 10 to 15 minutes of undivided attention centered on your brand for way cheaper. Then you can get the viewers to go watch more videos and interact with your brand in other ways if you're really smart about it. It's a no-brainer. You're quite literally insane if you're not capitalizing on the opportunity here.

Business execs think they know more than the YouTuber, especially if the YouTuber is very young. They want to control their integration, giving YouTubers a list of talking points to read verbatim with a call to action at the end. While this works to an extent, brands would convert exponentially more and see greater brand awareness and exposure if they let the YouTubers be themselves. A brand *should* say, "Here's a bunch of money, do what you do to connect with your viewers . . . as you've done on literally hundreds of videos with hundreds of millions of views. Do what you think would work best to get our brand talked about or our product sold. We trust you."

Brands who give the creative license to YouTubers and let them talk naturally like they normally would with their audience are getting more brand exposure and ROI than they could in any other way, guaranteed. And they should quit focusing on views but on what kind of impression they left. Will they remember you? Or will they hear another sleazy sales pitch?

Let me give you an example of the power of a YouTuber. I created an app called Finger On the App. I announced on my Instagram, Twitter, and SnapChat that everyone who downloaded and put their "finger on the app" at a certain time would have the chance to win $25,000. All you had to do was be the last person with your finger still on the app. The app was only available on iPhones and only in America. Even with these limitations, and without even trying, I had 1.5 million people install the app. I hadn't even promoted it on my YouTube channel, which at the time had more than 37 million subscribers. In contrast, a newcomer to the mobile video world, Quibi, launched around the same time as my app, but they had a $1.7 billion budget. Guess how many installs they got? Only 300,000 on the first day.

Finger On the App was inherently cool, so people talked about it. It naturally went viral. This is how businesses should pitch things. Change your mentality from wanting to get the most views to wanting to be the most talked about. My app was trending everywhere. I created a social footprint. And it all happened because I owned the app myself and had the control to spread the message my way. Brands come along and think they know better, but they would be so much more successful if they let YouTubers do what they do best. We know how to go viral. We know this digital world inherently because it's what we've lived and breathed our whole lives. We know how to get your brand talked about naturally. This is way more powerful than a pointless view.

YouTube isn't going anywhere. Google's parent company Alphabet and Google's Android operating system funnel so much crazy traffic to Google-owned YouTube. These companies have so much power and an unfathomable amount of cash that YouTube can't be a fad because of this. It's worth putting effort into something with such secure staying power. A lot of YouTube content is subpar now, but people will start to figure it out and it will be more competitive and expensive as time goes on. Figure it out now.

If you picked up this book, it's because you're already on YouTube or you are thinking about being on YouTube. Hopefully I've helped you understand that you absolutely should be on YouTube. You'd be nuts not to be. The opportunity to make money, grow a business, and spread a message is so huge, whether you're a regular person like me or a big brand. And this book has every element you'll need to do just that.

I first met Derral Eves in Dallas, Texas, for the sole purpose of talking about YouTube. I was working hard to achieve my dream of becoming the world's biggest YouTuber when I came across Derral's YouTube channel. I knew I had to meet him because he knew stuff I needed to know. So I messaged him, and I jumped on a plane. We've been YouTube data best buds together ever since.

Derral has pulled tens of billions of views on YouTube and even on other platforms because of his deep understanding of algorithms and the viewer. He owns VidSummit, the #1 YouTube data conference in the world. Only one or two other people in the entire world understand YouTube on the level that Derral does. I have one of the top three most-watched channels on YouTube, and I still talk about YouTube data and strategy with Derral because nobody is on the same wavelength as he is. Whether you're a creator or a brand, you can learn how to be successful on YouTube by reading Derral's book. Be open to what he can teach you, whether you have five subscribers or five million. Because remember, you should be improving at every level, always making better content. *The YouTube Formula* will help you do that.

MrBeast

INTRODUCTION

A magical ice-cream-pooping unicorn made me write this book. That's not where it started, but it's what pushed me over the edge. If you think this makes me sound a little crazy, let me explain.

It was 1999. I had recently graduated with a degree in advertising and marketing, and I had a stable job with benefits and potential for advancement. But then . . . I quit. It wasn't the right career path for me. I wanted to start my own business and create my own future. This felt crazy because my wife and I had our first baby, Ellie, that year, and I felt the pressure to provide for my family. Keeping a steady job and building savings would have been the safe thing to do, but instead, I spent all our money on the latest Macintosh computer and the software it needed (it would have been cheaper to go with a PC! Dang you, Steve Jobs). I was determined to be a successful business owner and ready to take on the world . . . from my kitchen table workstation at home.

I spent days coming up with my company info, designing a website, and making my own business card. I just knew I would be a great entrepreneur. But after only two weeks, my wife, Carolyn, let me know that my office needed to move from the family table where she fed the baby and where there were spills aplenty. She had gotten a degree in accounting and had a really good job and had put me through school, but we both wanted her to be able to stay home with Ellie and our future children. Carolyn was my business partner—she did payroll, invoicing, and bookkeeping. We discussed where the company was going and how we could get money coming in. I needed to find clients to start bringing in the money so Carolyn wouldn't have to go back to work.

My first order of business, though, was to move my work space. This took me to Staples to buy a desk for my computer. I was admiring all the things I wanted for my new business, walking down the aisle with color laser printers. If only I could convince my business partner-wife that I needed to buy one of these beauties! A fancy printer could help me show future clients work samples for ads I could make for them. My tech-geek daydream was interrupted by a man asking me a question about printers. He introduced himself as Chuck and was wondering if I could help him with picking a color printer. I explained the difference between the two printers in question because I knew all their specs—these were printers I wanted to buy for my business, too. I didn't have the money for a printer because I didn't have any clients. I recommended the more expensive printer because of its print capabilities and less expensive toner. He said, "I've been coming to Staples for years, and you are the best employee I've ever talked to." I informed him that I didn't work for Staples (which surprised him because I was wearing a red polo shirt and khakis—the classic Staples employee outfit). "Are you sure?" he asked. I convinced him by telling my story about starting my own business doing graphic design and websites. Wouldn't you know, he had recently started his own company called 1001 Business Cards and was looking for a designer. He hired me and my company on the spot to design his website and business cards. How serendipitous! I walked out of that Staples with a desk for $200 and a check for $300 from my first client, a $100 net profit. "That was easy!" in true Staples fashion. Owning a business was going to be a walk in the park.

I couldn't wait to go home and share my first success with Carolyn. My confidence soared and I knew I could do anything I set my mind to. However, only a few months later, my confidence had morphed into a feeling of uncertainty. I was going nowhere. I was barely making enough money to pay some of the bills and feed my family. We were dipping too much into our meager saving account. I needed more clients and fast.

Do You Know What Your Problem Is?

I went to see my dad to ask his advice. My dad had owned several businesses and done a lot of real estate ventures, and he had been successful at both. He gave me some simple advice, but it was some of the best advice I've ever received. He said, "You need to find a way to make more money with less time and with diverse streams of income." That was his standard way: he would always restate my problem before giving me fatherly advice. "Derral, don't think of solutions, think of the problem. When you focus and obsess on the problem, the solution will present itself." Take a minute and let that soak in. Einstein believed that our ability to identify problems was in direct proportion to the quality of the solutions we generate. He said, "If I had an hour to solve a problem I'd spend 55 minutes thinking about the problem and five minutes thinking about solutions." I believed Einstein's words, but I really trusted my dad.

I was determined to figure out what I was supposed to do by focusing on my problem, which was this: I needed more clients. I took the next morning to internalize what my dad had said and identified the problems that were keeping me from diversifying my income. I made a list of the problems associated with having only one client. I thought of every problem I might encounter that next week, month, and year if I continued on the path I was on. When I combed through my list, the solution presented itself to me. I had a list of businesses and all of their contact information sitting in a pile of business cards on my desk.

As an independent graphic designer, I created new business cards for my only client, Chuck, at 1001 Business Cards. He paid me $10 a card. Normally, I would have had zero contact with those businesses ordering cards. Chuck would get a new client, collect their logo, their picture, and their design requirements, and send them to me. I would design their card and email it to Chuck. After client approval, Chuck would have the cards printed and delivered. There was no reason why I would have interacted with these businesses. But now I had a plan to diversify. I wanted to sell them a five-page website design package for $299. I started cold calling these businesses with my offer. After spending a week calling every business card that I had designed, there were only two takers out of 200 calls. Selling over the phone was difficult; telemarketing wasn't for me.

I needed a new plan, so I refocused on the problem. I analyzed my talking points, learning from my telemarketing experience. I devised a new strategy that would take the process from a cold call to a warm lead, but it meant I needed to meet with these people in person. I asked Chuck if I could hand deliver the completed business cards to these businesses and offer my further services for ads and websites. He gave me the thumbs up and go-ahead, which landed me the opportunity to meet face-to-face with people who could be potential clients for my own business. These were the perfect candidates for businesses that needed a web designer.

When I'd present the finished cards to the owner or manager of the business, they were always impressed with the quality of the card. We didn't do simple cards; we printed in quality full-color graphics with all the bells and whistles of a premium business card, which was uncommon then. As soon as I saw the moment of excitement or approval of the product, I'd pounce. Here was my window to get a new client! I told them I was the card designer, I was glad they liked my work, and I could do a lot more for their business by helping them create a website. I told them that every business needed a website—this was the future in marketing because leads and sales were going digital. It actually was a tough sell, because most businesses in my area were still on dial-up Internet at the time. If you are unfamiliar with dial-up Internet, let me explain it this way for you youngsters: dial-up Internet ran at 56/K (kilobits) per second. Not megabits. Which means that if you wanted to download today's latest update of *Fortnite*, it would take you 30 days, 22 hours, and 3 minutes, assuming your connection wasn't interrupted and you didn't have to start all over again.

You could imagine that people didn't see the value in spending money on something so slow and widely unused, but I would start using my sales pitch. It really was going to be the way of the future. Even if only 10% of the city was using the Internet to look for a service, their business was guaranteed to take that call if their competitors weren't online yet. Having a website would give them a place to host frequently asked questions, share a bio about the business and/or the owner, and, most importantly, generate leads. It was a digital sales brochure and the future of marketing.

This new plan worked almost every time. I was converting nearly every business card client into a website client. I got more than 100 clients originally, plus referrals at every turn. And as a bonus, they would need a second business card printed because the first one I designed hadn't had a website listed on it! Score! I sold these updated business cards for $20, helping my original client, Chuck, and giving him additional clientele. He had been my first and only client, and I became his biggest client over the years, bringing in lots of business for him.

Your Client's Success Is Your Success

But website design was one-off work. I only got paid once by each client, which meant that I was always hustling and trying to make a sale. There were a lot of ups and downs, and it was starting to wear on me. It was time to visit Dad again. I told my dad of my predicament, and he smiled as he said, "Remember the advice I gave you?" Of course, I did. I had gone from one client to hundreds of clients, even getting other designers to help me so I could save time while bringing in more money. Dear old Dad smiled and said, "Well, I only gave you half of the advice. You needed to learn that first part before you were ready to learn the next. You've been banging your head for some months now, right? There are a lot of ups and downs?" It was like he was reading my mind. "You've been losing sleep, too, haven't you?" How did he know? "Now you are ready to understand the second part: Success is all about acquisition and retention of clients and having them come back wanting more. You got the acquisition down. But how is your retention? How many clients pay you monthly?" I explained the one-and-done nature of website design. Every single client paid me once. Dad said, "You'll always have ups and downs and sleeplessness without retention. Figure out retention—how you can keep them wanting more, and you can figure out success." If you haven't realized it yet, my dad is a genius.

So I analyzed my problems again. What could I do differently? I came to the conclusion that there were two ways I could get recurring monthly payments from clients: hosting and ranking, that is, Internet marketing. All websites needed a host, which required monthly fees. So I started a hosting company called FatBoy Hosting. The name referred to the size of the hosting packages, not my personal weight (or did it . . .?). We had everything from small to XXXL to Blimpie-sized packages. I also knew that companies paid monthly for Internet marketing. They needed to have websites get ranked in directories and other services. So I started doing website ranking for companies that paid me monthly. My business soon stabilized and started producing residual income. And I got more sleep. Thanks, Dad.

I continued to build websites for clients, hosted their website, and helped them get ranked on the relatively new Internet, or the World Wide Web as it was more commonly called then. I used search directories like Yahoo, Excite, Ask Jeeves, and AltaVista. "Google" wasn't yet a household name. I did this for several years, trying to expand my company. In 2005, I had hired a few new employees and needed to purchase some inexpensive office furniture, so I went to Craigslist to look for a good bargain. When you're fresh out of college and excited about a glowing future in entrepreneurship, you buy your office desk brand new at Staples, but when you've been around the block a few times and the shine has worn down, you save money and buy secondhand goods (or even better, get them for free) that work just the same!

While I was on Craigslist, I saw a listing for a contest to win a free new iPod Nano for anyone who could get people to join a new website called YouTube. The previous iPod on the market was the size of a brick and held a weighty 1,000 songs! Plus, if you threw it at someone you might kill them because it was heavy like a brick. Steve Jobs had just announced this new iPod Nano, which was the size of a pack of gum (much lighter than a brick) and held even more songs. I wanted that iPod! I signed up for YouTube, spammed all my contacts and clients to do the same, and even created new emails for myself personally to increase my chances. I became one of the lucky winners of the latest and greatest iPod on the market.

But then I started watching videos on YouTube and was blown away by the quality of the videos and the lack of the dreaded "buffering" load time so common then. I was hooked! I learned that you could embed videos from YouTube to any website, and people could play them wherever they were in the world. BAM, an idea came to me: I could upsell my 865 clients to embed a video on their website. It would be so easy, it would be like printing my own money. No one was doing this. I could be the first to put videos on my clients' websites.

In November 2006, YouTube was purchased by Google for $1.65 billion. Naturally, the platform's videos started to gain ranking traction on Google searches—now that they owned it, of course they wanted to help it get noticed. At this time, my job was to keep my clients' websites ranked as #1 on the front page of Google. But there was an anti-spam czar at Google named Matt Cutts who made my life a living hell. Matt Cutts was my nemesis. My whole day was spent trying to figure out hacks to game the system and get my clients' websites highly ranked, and Matt and his team would find exploits to the system and shut them down every time. I felt like I was stuck on a roller coaster with no end in sight.

I was sick of fighting with Matt Cutts, and I was sick of getting the dreaded phone calls from my unsatisfied clients every time their ranking dropped. This was driving me insane. So I focused on the problem again. I looked at my business and asked myself what would be the easiest way to accomplish what I needed to do. I had an epiphany: if I stayed in line with Google's goals, I would get ranked every time. I didn't have to fight the system (or Matt Cutts) anymore. It sounds so simple now, but it was a big aha moment for me at the time. I looked at everything that Google was trying to accomplish. I read every blog post and watched every talk by Google founders Larry Page and Sergey Brin, and Google CEO Eric Schmidt. I started to listen to what they really wanted. They talked a lot about organic ranking and tracking. They also talked about the future of Google, with artificial intelligence that would look for patterns to predict what people would want. I made sure I met all of the requirements when it came to what Google wanted.

It was then that I noticed some patterns from my clients' data. Remember the 865 videos I made and embedded on their websites? Most of them were on the front page of Google without any hack or help. WOW! You mean, I didn't have to fight Matt Cutts? And I didn't have to get frantic calls from clients? Sign me up! All I needed to do was make videos that were search-friendly. So I created a plan of attack. I made a series of videos for a few select businesses and focused on generating a lead. My brother-in-law is an optometrist, and I asked him to let me market his practice with videos. I was able to rank for some hard key terms that would have taken a lot of effort for a website but showed up easily in the results in a matter of hours when I used videos. *Crazy!* To make sure this wasn't a fluke, I tried it with a second business, a personal injury attorney. We got the same results: a high Google ranking and the phone ringing with leads.

I had done it; I had solved one of the biggest problems for my company, and I wanted to go all in on video and YouTube. I needed to become the expert helping businesses by generating leads and sales with video. I kept my big clients who were paying thousands of dollars a month, but I sold everything else. It was my point of no return. Like when Spanish commander Cortés sank his ships and conquered the Aztecs. In sinking the ships, he gave himself one option: succeed or die. There would be no means to turn back. Likewise, I sold all my website and Internet marketing clients to an SEO company in Salt Lake City, with the clause that I could keep all the video marketing work. They agreed because they didn't know what I knew, that videos ranked in Google without any blackhat hack or strategy. They were being promoted because YouTube was owned by Google. Google wanted videos to be found, so they moved more users to YouTube. So I shifted my focus from website creation and marketing to video creation and marketing . . . and I watched my clients' rankings and revenue skyrocket because these videos were ranking on page one of Google.

By this time, I was itching with excitement for what I was witnessing in the evolution of marketing: online video was going to be a big deal, and I wanted to be a part of it. Video power outweighed word power by an overwhelming margin. Viewers were much more likely to do what advertisers wanted them to do—pick up the phone and call, make a purchase, or sign up for a service—after watching a video. And these videos could be made by regular people who were getting incredible distribution to millions of viewers . . . using inexpensive camera equipment. This had never been done outside the realm of television studios and movie sets.

In my videos, I hyperfocused to dial in messaging with the goal of getting the phone to ring. It was so important to know everything about the person making the phone call. I would grill my brother-in-law and his office staff on all the things people would ask and what they needed help with. Then I would take those questions and turn them into talking points for the video with clear solutions on how we could help. I asked a ton of questions to get a good handle on the niche and what would work. Then I would make 10 solid videos that would show up in search. Once the messaging was working—the video was ranking and the phone was ringing—I would go to another city and do it all over again. (I only worked with one business per niche in each city.)

I had started with my optometrist brother-in-law, but I replicated my work in that niche hundreds of times in hundreds of locations. When I got a system down for one niche, I could rinse and repeat for any client in that niche. Why re-create the wheel when I could slap a new logo on what was already working? I had my systems dialed in so well. The clients were happy, which meant that I was happy. Matt Cutts was no longer the enemy; he was my best buddy! Talk about a win-win.

Uniting People around Their Passion

At that time, my relationship with YouTube consisted of creating videos and uploading them to YouTube for the sole purpose of getting a Google ranking for lead generation. Little did I know that this was a fraction of the potential of YouTube. There was a huge opportunity to reach millions around the world, and I was about to witness this with my next client.

One day I got a phone call from one of my clients, Wade Beatty, who had a lead for me. Wade owned a local pest control company, and I had done some successful video marketing for him. He had gone to spray a piano store for cockroaches, and the store owner asked if he knew anybody who did YouTube, websites, and marketing. Wade said, "You have to talk to Derral." So I talked with the store owner, Paul Anderson, who showed me the amazing videos they were making to try to sell expensive pianos. They put grand pianos in beautiful locations outside—on top of mountains, in the desert, in a forest—and the music videos were amazing.

I asked Paul what his goals were with his store "ThePianoGuys." He told me he wanted to make enough money from his YouTube videos that he didn't have to work at the store anymore. He didn't actually want to sell pianos, he wanted to make awesome videos that people would love. With the talent of Jon Schmidt at the piano, Steven Sharp Nelson on the cello, Al van der Beek producing the music, and Tel Stewart doing the videography and editing, they had the recipe for success. Their musical passion turned into a channel, and the channel turned into a business that was making so much money that they didn't need the store anymore.

When ThePianoGuys' YouTube channel exploded, I was able to be a part of the true power of YouTube for the first time. They had had minimal subscribers when we started, and they had grown to more than 1.8 million subscribers in 12 months and had over a hundred million video views. They had developed a passionate, dedicated audience. (They never sold a single piano! But they did sell out stadiums worldwide.) This success shifted my focus 100% to working with creators to help them build YouTube audiences, and I never looked back. Yes, I sank my ships again, and I haven't regretted it for a second.

Uniting people around a similar passion was exhilarating. It was an intense realization that I could help people find their own dedicated global audiences. It didn't have to be confined to a geographical location like my previous work had been, and it didn't have to be all about business. In fact, it *had* to be about more than the business. You could find passionate followers in every corner of the Internet and bring them together around your content. This was about more than generating leads. It was the perfect crossover between money and passion, something I had been missing all those years leading up to this moment. I had learned a lot about algorithms and rankings and the mechanisms that worked, but now I could see the other side of the coin. Creating content to inspire, educate, or entertain was the missing link. Everything I had learned about algorithms, people, and messaging came together in this new moment of clarity. Audience development was my thing. I realized I was really good at creating a community around content, so I dedicated my career to learning what makes an audience click.

To date, I have created and developed a plan and content strategy for 25 different YouTube channels. I have helped them grow from zero subscribers to more than a million subscribers each. With my formula, we've generated more than 59 billion video views in total. I've seen this change so many people's lives: they not only become full-time YouTube creators, but they build sustainable businesses and brands.

For businesses, this can be a game changer for your bottom line. Here's where the magical ice-cream-pooping unicorn comes in. I was executive producer on one of the most viral video ads of all time, and it was for a step-stool called Squatty Potty. It showcased this magical unicorn demonstrating the advantages to using the toilet stool while "pooping" ice cream. Yes, the stool is a real product (unfortunately, the unicorn is not), and it did more than $45 million in sales in one year. That's a powerful ad! After I watched its success, I knew it was time to share what I know about YouTube so that others could learn how to tap into this resource as well.

Don't hesitate to jump in with both feet as I teach you all things YouTube: the ins and outs of the platform, the opportunities waiting for you, and the formula for doing it right. You might even have to throw out some things you thought were true about YouTube, but no need to worry; you'll relearn quickly and have the solutions on hand. Reach for this book as a reference as you move forward to create amazing content and get the results you've always wanted.

I've been on YouTube since 2005 and have seen how it has evolved and how it can change a business, and more importantly, how it can change people's lives. YouTube is one of the most powerful platforms we have to connect with the world and grow an audience using video.

I can't wait to show you how to harness the power of YouTube . . . let's get started!

PART I The Platform

1 Try, Fail, Analyze, Adjust: A YouTube History Lesson

Have you ever watched the first video uploaded to YouTube?

It's an important question. No really, it matters. I don't think it matters just because I'm a YouTube expert and think it's high-quality content (spoiler alert: it's not). I think it matters because it's history, and we can learn so much from history. As a video guy, I love documentaries and biopics. They fascinate me because learning about the past helps us understand and navigate our world in the present. When we explore history, we see how decisions and events affect people, whether for good or bad, and how they impact people's families, communities, and ultimately, the entire world.

But why the heck am I talking about the impact of history on the world in a book that's supposed to be about YouTube and making you money? I'll tell you why: studying the history of anything can benefit someone who wants to learn more about that topic and succeed in that space. It should be obvious that the same applies to YouTube creators as well. I truly believe that if creators and businesses will take the time to learn from YouTube's history—how it became the mega-platform that it is based on decisions that lead to failure and success—they will become better content creators and businesses and be more equipped to generate their own success on the platform.

So how did YouTube start, and how does that affect you and your content?

In July 2002, the prestigious start-up company PayPal had just been acquired by eBay for $1.5 billion. This created a lot of buzz in Silicon Valley. Ideas were being tossed around for websites, apps, and platforms that could possibly bring in a lot of money and transform the world into a digital money-making giant. Three PayPal employees, Jawed Karim, Chad Hurley, and Steven Chen, were some of these idea generators. They soon came up with the idea for the YouTube website, but it was nothing like the website we are familiar with today—they started it as a dating website.

From their makeshift office in a garage, the domain name "YouTube.com" was activated on February 14, 2005, Valentine's Day—the perfect day of the year to start a dating website. On April 23, they uploaded YouTube's first video called "Me at the zoo." It was 19 seconds of Karim at the San Diego Zoo talking about the elephants. If you go and watch it now, you'll laugh because you know he was trying to get the website to be a place to find a date, and he makes an innuendo about the anatomy of the elephant. The video was actually pretty good quality for nonprofessionals at the time.

Now that they had the ball rolling and the servers going, these guys needed active website users. Who was the demographic of people looking for love? College-age students. So they pitched at the nearest university campus, Stanford, and canvased pretty heavily, passing out flyers to everyone who would take one. Their slogan was, "Tune in, hook up." There were virtually no videos on the site yet, so they uploaded footage of 747 planes taking off and landing. There was no rhyme or reason; they just wanted videos on there. "The whole thing didn't make any sense," Karim said. "We were so desperate for some actual dating videos, whatever that even means, that we turned to the website any desperate person would turn to: Craigslist." They ran the ad in Los Angeles and Las Vegas, offering to pay women $20/video to upload videos of themselves. They got exactly zero responses.

Here is where it gets interesting and applicable: Once people began using their "dating" website, the trio looked at the data coming in those first few weeks and months, and they realized that the website's handful of users were not coming for dating at all; they were coming for self-broadcasting. They were posting videos of themselves and their friends doing funny or embarrassing or weird things. They were posting videos of their pets, videos of snowboarding, videos of random places and things, and the like.

At this critical juncture, Karim, Hurley, and Chen had a decision to make: should they continue to push YouTube as a dating website as planned, or should they change their business model entirely because the data showed that the usership was not the I-want-a-date crowd? "Forget the dating aspect," said Chen. "Let's just open it up to any video." Herein the power of YouTube was born. Based on the data feedback, they switched gears and catered to what the users wanted.

In June, they created tools that encouraged self-broadcasting. They supplied a growing ecosystem of whatever random videos were being uploaded by the people. They launched an "embed" video option that became a game changer for websites and promotion. In short, they gave self-video creators the platform and the control to share their videos with the world from anywhere in the world, because that's what the people wanted. These website users were not looking to hook up on YouTube; they were finding a place to put their work and their creativity.

That decision to pivot led to YouTube becoming the most powerful video platform in the world and disrupted the entertainment industry as we knew it. It was a game changer in video entertainment, taking creation from the hands of the few to the hands of everyone, if they so desired. With access to a recording device and an Internet connection, anyone could broadcast a video to the entire world! This is the norm today, but think about how monumental it was in the beginning. Big businesses and brands started to pay attention, changing their content creation and advertising strategies. With power shifting to the regular joes, brands seized on the opportunity to sponsor creators who had a unique, organic viewership. Big business had never had this kind of competition before, and it was a force to be reckoned with.

The website continued to grow quickly. Google saw the early potential of the site, and they acquired YouTube in 2006 for $1.65 billion. Today, more than one-third of all Internet mobile traffic comes from YouTube traffic. There are more than a billion combined hours watched on YouTube every day, and almost two billion logged-in users visit the site every month. Nearly 100 countries have local versions of the platform available to them.

Do you think all of this would have happened if the guys had ignored the data feedback, deciding to stick to their original plan, and insisting that YouTube had to be a dating website? They had tried the dating website, and it had failed. So they focused on the problem, analyzed what was working and why, and they adjusted their strategies to support more of that.

YouTube's origin story is the ultimate meta-example of how to try, fail, analyze, and adjust to succeed on YouTube. This formula *is* the YouTube Formula. Understanding its history will help you as a creator or business understand how to utilize the formula for your own success. You have to analyze what's working and what's not, and make changes accordingly. This is the premise on which the whole book is built. If you can grasp this "big idea" foundational formula, you're starting out on the right foot, and you're ready to learn the step-by-step tweaks that make all the difference in the wide world of YouTube.

In Part I, I break down the algorithm so you know exactly how the YouTube platform works in order to become a part of it. In Part II, I open your eyes to the endless opportunities available on YouTube—opportunities for exposure, artistry, collaboration, sponsorship, merchandising, and business ownership. I tell you how all different kinds of creators and businesses have seized on these opportunities and gone beyond "making a living" from YouTube. There is so much money to be made in so many ways on YouTube, but even more, there is so much power in influence. I can show you how your influence can make a big difference.

In Part III, I dissect the YouTube Formula for content planning, creation, execution, distribution, analysis, and adjustment. I teach you how to find your audience, speak to them, and convert them into your own loyal community. I teach you the importance of traffic sources so you know where the viewers come from and how you can get your content seen. I help you read metrics graphs so you can recognize data patterns. Your YouTube success depends on developing these skills, so get ready to learn and embrace them.

I've helped countless YouTube channels tap into growth opportunities they could not see on their own. And I've helped creators and brands learn the steps to get views, make money, and build businesses. If you follow the Formula and open your mind to the opportunities I am going to show you, you can get the results you've always wanted on YouTube.

2 The YouTube Ecosystem

In order for us to understand how YouTube really works, first we need to look at how it operates as a digital ecosystem. A digital ecosystem works much like a natural ecosystem: there are a lot of moving parts, and all of those parts affect the organization as a whole. Grade-school science taught us about energy flow in a natural ecosystem; photosynthesis, plants and animals, decomposition, and nutrient conversion are all part of the cycle. Every factor in the chain has its job to do, and if it doesn't work right, it affects the entire operation.

YouTube's ecosystem also has a flow and cycle, and its contributors affect the whole, for better or for worse. This digital ecosystem includes the creator, the viewer, the advertiser/brand, copyright holders, multichannel networks (MCNs), and YouTube itself.

Here's a quick summary of how the YouTube ecosystem works: creators make videos and upload them to YouTube. Brands pay YouTube to run advertising alongside uploaded content, either before or during a video. When a channel meets the ad sharing program requirements, it gets a cut of the money from the ads running on their content. Brands also connect with creators who they think will be able to increase brand awareness and/or their bottom line. This influencer marketing is a huge part of YouTube's ecosystem. The viewers come to interact with content, creators, and communities. They watch, subscribe, comment, like and dislike, save, and share. YouTube as a website is the host of the ecosystem, but as a company, it's a part of the ecosystem. YouTube the company has to make sure everyone in the ecosystem is satisfied. They field complaints and legalities. They ultimately make the rules, but the rules evolve over time based on feedback from the ecosystem and what needs to be addressed. MCNs played an important role in the beginning of YouTube, connecting brands with creators and managing other elements of the creator experience. They also helped try to problem-solve because YouTube didn't have the creator support at the time. Creators don't have to work with MCNs; they can manage their own channels and deal directly with brands or work with agencies to connect with brands. Finally, copyright holders want their original work attributed to them without being stolen or copied. They want to keep any financial benefit from that content coming back to their original content.

To be a part of this digital ecosystem, understand the role each contributor plays, especially the role you intend to occupy. For example, if you are a creator, become familiar with YouTube's guidelines so your content can get monetized and stay monetized. Don't steal or copy someone else's content, but if you intend to use clips of scenes or songs or any other copyrighted material, know how to go about it legally. Your content's success depends on your understanding your role in the ecosystem. Your YouTube experience should be more than starting a channel and uploading videos. In fact, if this is your methodology, your video will never reach viewers. YouTube rewards original content that's made for a specific audience, so if you learn the system and follow the rules, your videos have a greater chance of being seen.

Beware of Copyright

The first consideration in the ecosystem should be the viewer. If nobody is coming around to watch, the rest of the ecosystem is dead in the water. When YouTube was new, the viewers were a fairly specific demographic because the content was fairly specific. People were uploading personal videos to share with their friends and family, so there was an audience there, but where the viewership really had the potential to grow pushed in the entertainment direction. People were uploading clips from TV shows, movies, comedy bits, and the like. Viewers were also coming to find clips from popular culture and the news. It was easy to upload this stuff, and it was easy for the viewers to find.

From 2005 to early 2007, YouTube users had been uploading content generally unregulated and unsupervised. This included a lot of original content, sure, but it also included protected content that had been created by another person or company. Obviously, this was a direct infringement to the copyright holder of that content.

It's important to note here that YouTube users weren't doing this sneakily or maliciously. They just wanted to share things they loved, and it was so easy to do. Do you remember Napster from the turn of the century? For those of you born in the 2000s, let me tell you a story.

Imagine a world where you couldn't listen to your favorite music on demand. The only way to hear your favorite song was to sit by the radio all day, waiting. If you wanted to listen on demand, you had to buy the entire album that included the one song you wanted to hear. Then along came Napster. Napster was the original widespread file-sharing platform. Like, *the* pioneer of all digital media sharing on the Internet. Audio files, mostly songs, were shared as MP3 format files, and anyone could download any file for free. FREE! This was huge for music fans all over the world—the people loved Napster. Who wouldn't want unlimited access to their favorite bands for exactly zero dollars? Well, I guess not quite everyone was a Napster fan . . . namely, anyone who should be making money from music sales.

If you don't know the rest of the story, I bet you can guess what happened next. Lawsuits. Shut down. No surprises here. Actually, just before Napster was created, Bill Clinton, then-president of the United States, had signed the Digital Millennium Copyright Act (DMCA) into law in 1998. The DMCA has regulated digital copyright issues and reinforced offender penalties ever since. (However, websites hosted outside the US are regulated by the United Nations World Intellectual Property Organization (WIPO).) What happened with Napster set the stage for media sharing regulation from that point forward.

YouTube could have looked at companies like Napster as examples of what not to do, and implemented their Content ID system from the time they opened up shop in 2005, but they didn't. In March 2007, a little company called Viacom, along with several others, sued Google and YouTube for $1 billion worth of copyright infringement issues. Reuters reported that YouTube was only taking copyrighted content off the site after a copyright owner had requested it, but there was nothing being done on the front end to prevent that content from being uploaded to begin with. Further, the lawsuit claimed that YouTube knowingly let this happen because they were making money on all that content.

The Content ID system wouldn't be implemented until 2007, the beta in June and full rollout in December. The ID system would attach a unique "digital fingerprint" to new uploaded content. Content could then be tracked and measured against already existing copyrighted content so YouTube could catch infractions.

Now that the law had been involved, YouTube had to settle some things for copyright holders. First, they literally had to settle that hefty lawsuit (the terms of settlement were undisclosed), but they also had to settle how they would proceed from there, and the Content ID system was the answer. I can't stress enough how monumental this was to YouTube's success. YouTube likely would have followed the fate of the original Napster and been shut down if the Content ID system hadn't been implemented, and a lot of creators and businesses like you and me would be doing something else. It was a total game changer.

Ad Revenue Sharing

In 2007, while the big copyright problem was simmering down, YouTube added two features that would make a similarly huge impact on its future. These features would be problematic in some ways, but also they would significantly change advertising and how creators could make a living from YouTube. Which features could be this consequential? (1) In-video advertisements and (2) the Partner Program.

Advertisements on YouTube have evolved a lot over time. They used to be display ads or they appeared underneath the content, but with these changes, ads would pop up right in the content where the viewer was looking. And now the creator of that content could be compensated via ad revenue sharing. This is how the YouTube Partner Program began. Creators became super motivated to make good content that would get more viewers watching, because now, more views equaled more money. Creators and businesses really wanted to be YouTube Partners!

Unfortunately, this also meant that creators realized they could use tactics to get people to click on their videos, even if those tactics were divertive. Their aggressive "bait-and-switch" strategies included misleading titles, sensational thumbnails, and superficial content that strayed from the original purpose of the video. They wanted to get people to click on their videos at any cost so they could produce revenue from the ads being integrated with their content.

Let's not forget the first member in the YouTube ecosystem here: the viewer. This new ad sharing program created a big problem with viewer satisfaction. People began spending less time on YouTube because of the in-your-face ads and because of clickbaiting that hinted at what they were looking for but actually didn't satisfy that end at all in the content itself. In short, viewers felt tricked and unsatisfied.

In addition, many YouTube viewers got angry at these "sell-out" creators who ran ads in conjunction with their content. They even went so far as to join channel boycott movements against creators. Seeing ads today is just part of the online experience, but back then, it was such a big deal that it interrupted the ecosystem.

As you can see, the integration of the Partner Program further complicated YouTube's delicate ecosystem. Keeping advertisers happy was an obvious priority because that's where the money came from, but YouTube also had to keep creators and viewers happy to achieve this, and the task was proving to be extremely difficult. Creators wanted their fair share in ad revenue without being labeled a sell-out, and viewers wanted to watch content without feeling tricked or sitting through ads on every video.

Let's pause here for a minute. Before we run away with all the problems the Partner Program created, I want to emphasize how monumental its existence was. Google had pioneered advertisement revenue sharing with its AdSense program, and they implemented that program with YouTube. They took it to another level going from display ads to video ads, because video ads were more effective. They could charge advertisers more money for video ads, so YouTube made more money this way. Because YouTube didn't create their own website content, they had to incentivize creators to make good content that would get people coming to the platform.

AdSense was the origin of advertisement revenue sharing. This was absolutely uncharted territory! It was like the great California gold rush, but it was the gold rush for digital marketing in the twenty-first century. Companies had never offered a portion of their revenue to the general public before!

YouTube said, more or less, "Hey folks, if you make good content that viewers will come and watch on our website, we'll share some of the advertisement money we get with you."

And creators were like, "Wait, really? You mean I can get compensated for my hobby? Potentially earning enough that I can replace my boring nine-to-five income doing something I actually like, and can make more money doing? Well, then I'm going to make the best darn videos you ever saw!" And the gold rush was on. The channel owners grabbed their pickaxes, installed sluice gates, and started panning for gold.

So many creators, businesses, and advertisers saw the potential for massive payout with ad sharing, and they wanted in. The risk was that they might not actually make any money if they didn't get viewers, but the opportunity was worth the risk, and it paid off big for so many of them. This was a genius move by YouTube! They were enlisting a worldwide army of creators to do the grunt work to get visitors to their website for a piece of the revenue. Channel owners were ready to compete against each other for a chunk of the gold.

Advertising companies could get into this market very inexpensively. They could track their market, get a lot of eyes on their product quickly without expensive campaigns, and they could do it for literally a fraction of traditional marketing cost. A lot of big brands and businesses gave it a snooty pass because they just didn't realize what they were missing in the beginning. They were too good for it.

One of my favorite examples of a YouTube advertising success story comes from a product called Orabrush. Orabrush is a tongue cleaner that was invented in the early 2000s by a guy named Robert Wagstaff, aka "Dr. Bob." Dr. Bob had tried to market his tongue cleaner via traditional product-pitching means, but the companies he approached wanted nothing to do with it. He even invested a lot of his personal money to run an infomercial. It flopped. So at the ripe old age of 75, Dr. Bob turned to a marketing class at his local university and asked if anybody had any bright ideas.

My good friend Jeffrey Harmon was a student in that class, and he told Dr. Bob that he thought they could sell the product online with a YouTube video. He took on the tongue cleaner project with the promise of Dr. Bob's personal motorcycle as payment for the campaign. Jeffrey and some creative friends made their YouTube video for just a few hundred dollars, and it went viral. People wanted to know where they could buy Orabrush in their own locations, and distributors started to pay attention.

An ad campaign like this had never been done on YouTube, but Jeffrey saw its potential. "We took a product from zero sales anywhere to worldwide distribution," Jeffrey told me. "And we didn't do it with traditional marketing. It was 100% YouTube. We couldn't have made it happen any other way." Orabrush became a multimillion-dollar brand that is sold in more than two dozen countries in 30,000-plus stores.

For the record, Orabrush's YouTube channel has more than 38 million video views to date. For a tongue brush. Orabrush was acquired by DenTek in 2015.

YouTube leveled the playing field for a small-town, old-man inventor and a few college kids. They had access to the market that was simply unavailable to regular people before YouTube. It literally changed the trajectory of their lives. Jeffrey Harmon became cofounder and chief marketing officer at Orabrush and has gone on to cofound the Harmon Brothers marketing agency with his three brothers, where they have created extremely successful online campaigns for PooPourri, Squatty Potty, Purple, Lume, and many more businesses. Others who worked on the original campaign also have made successful careers out of the path that was set from Orabrush's beginning.

Another friend of mine, Shay Carl Butler, began his YouTube career way back in 2006 and was one of the original Partners, so if anyone has a good handle on how the program began and how it has changed ever since, it would be him. "The YouTube Partner Program was so exciting in the beginning, and everyone who knew about it wanted a piece," Shay Carl said. "YouTube had a lot of problems to work out. I remember when it seemed like everybody was mad at some point: viewers were mad at creators, creators were mad at creators, viewers and creators were mad at YouTube, and so on. But YouTube has done a good job overall of working out the kinks." Shay Carl's personal channel is where he began in 2006, but he started his family channel, *Shaytards*, in 2008, and it has become the main YouTube lifeline for his family with around five million subscribers to date. We'll go into more detail about the Partner Program in Chapter 6.

Because of this "digital gold rush" and the amount of creators and advertisers it brought in, YouTube was not equipped to respond to the masses. This is where the multichannel networks came in. MCNs offered to be the go-between for other contributors in the YouTube ecosystem in exchange for a piece of the profit. They helped creators and businesses with audience growth, resources for production, and brand opportunities. They matched advertisers with channels that suited their particular products or services. They dealt with rights management. They gave YouTube some breathing room to worry about other things. MCNs have had both good press and bad, but they did alleviate a lot of YouTube's headaches in those formative years of ad revenue sharing.

I recently sat down with Jim Louderback, CEO of VidCon and former CEO of MCN company Revision3, and talked with him

about multichannel networks on my podcast *Creative Disruption.*
He talked about how MCNs affected the early years of YouTube's
ad revenue explosion. We discussed the ways they helped, but also
the problems they created. "In the end," said Jim, "a lot of MCNs
did not provide the value they offered. They brought on too many
creators and brands to manage, and there wasn't enough revenue to
go around."

Once YouTube had a better handle on operations, they offered
their own Partner support rather than losing creators to outside
MCNs. In 2011, YouTube acquired Next New Networks, a company
that had been managing a lot of YouTube's early creators. YouTube
was ready to take back control internally and put that money back
into their own pocket. For this reason and others—like fewer creators
signing up, and a smaller margin of revenue per view—MCNs have
seen a significant decline on YouTube in recent years.

An Evolving, Thriving System

As you can see, ad revenue sharing completely changed the YouTube
ecosystem. YouTube had been a minefield of instability in its fledg-
ling years. They were learning the hard way that their ecosystem was
a delicate balancing act—among its copyright holders, its viewers, its
Partners, and its advertisers. There has been an "Adpocalypse," countless issues with the algorithm, FTC COPPA children's privacy issues,
Adpocalypse 2.0, and more. YouTube has learned to deal with the
problems and tried to make changes to satisfy the masses, but it is a
constant effort.

When we come full circle back to the viewer, the first compo-
nent in the ecosystem, we have to consider how much the viewer
has changed. YouTube really wants to have satisfied viewers. Over
the years, they have tried to modify their recommendation feature to

figure out exactly what each viewer might want to watch. They know that happy viewers will stay around longer, and viewers who stay around longer will produce happy content creators and happy advertisers. And the more the viewers watch, the more money everybody makes. The thousands of changes that have been made to the algorithm over the years have literally paid off, so the better the algorithm gets, the happier everyone will be. Get ready to dive deep into the algorithm in the next chapters.

YouTube must be figuring out some things, though, because they have seen 31% year-over-year growth! In 2020, YouTube announced their revenue for the first time ever. In 2019, they made $15.15 billion, which was nearly double the year before! That is mind-blowing, in both the amount of money and the percentage of growth. People watch more than five billion YouTube videos a day. *Billion.* To really grasp how much bigger a billion is than a million, consider this: one million seconds is roughly 11 days, while one billion seconds is 31½ *years*. Now give a second thought to that $15.15 billion figure for YouTube's 2019 revenue, and gasp. And they are really only just getting started.

YouTube began as a dating website for a handful of college co-eds in California in 2005. Now it reaches every corner of the globe on every device. Roughly one-third of the entire population of the earth is watching YouTube regularly. Again, we are talking in billions. Where it used to be a specific demographic, the YouTube viewer is now *everyone*.

Creators and businesses can proactively position themselves to win on YouTube by learning about their own role in its ecosystem and the mechanics of a good channel. These are nonnegotiables if you want to succeed on the platform. Don't try to game the system; try to align yourself with YouTube's goals so fewer problems come up and you can focus on good content creation. YouTube has changed exponentially since its inception, and its ecosystem has changed, too. If you want to be a part of that ecosystem, you have to understand how it works and how you can adapt to it, because it will continue to change. You can adapt intelligently by looking at the data YouTube gives you. I'll show you how the algorithm works and how to create and adjust from its data.

3 The YouTube AI: A Deep Learning Machine

A YouTube creator who recognizes the need to adjust to the data but doesn't have a clue how to do it is like a gardener who wants home-grown produce but has never planted a seed. Becoming a successful gardener doesn't happen overnight, and neither does becoming a YouTube pro. You have to grab a shovel and dig in. There will be blisters on your metaphoric hands in the beginning, but as you develop your data-digging muscles, you'll start to unbury a network of underground connections and discover a whole new world of the hows and whys of YouTube and what it takes to produce successful content.

YouTube's artificial intelligence (AI) is an evolving structure in the digital ecosystem, and it takes work to understand and utilize, because it's malleable. You'll need to be malleable, too, meaning you have to adapt your strategies according to what's currently working. Your best chance at doing this successfully depends on your knowledge of the systems at play.

The AI Evolution

As contemporary YouTube users, we have grown accustomed to the site dishing up what we like, unprompted, but it hasn't always been this way. Initially, YouTube primarily was a place to find answers to our questions, like how to change a tire, and a place to be entertained, like watching cats play keyboards or laughing at kid videos, like "Charlie bit my finger." It was built on a simpler system that wasn't good at making recommendations. But YouTube today has a complex machine learning system that has gotten really good at guessing what people want. Let's take a closer look at how its AI has changed over time and why that matters to you.

About 2011, YouTube started making system changes with one purpose in mind: get people to stay on the platform longer. A You-Tube researcher working on this issue found some gaping holes in the framework. For example, a huge portion of YouTube viewers had gone mobile by then, and YouTube didn't have an accurate system for tracking user behavior on mobile devices. Palm to face. There was work to be done.

Since July 2010, YouTube had been using a program called Leanback that queued up-next videos that were ready to load after the video being watched was over. There was an initial increase in views, but soon they plateaued. They got the same results from a follow-up AI program called Sibyl.

YouTube joined forces with Google Brain, Google's machine learning team, whose AI development and tools were leaps ahead of the field. Their goal was to build a system with the Google Brain foundation. Their main objective continued to be viewer longevity. On March 15, 2012, YouTube made the switch from a "View" algorithm that rewarded video view count to a "Watch time" algorithm that rewarded viewer duration. This AI followed the audience everywhere to ensure it found the right video to put in front of them.

It had the capability to recommend adjacent videos rather than clone videos ("adjacent" meaning similar but different enough to keep interest). "Clone" videos inevitably pushed viewers off the platform because they were watching basically the same thing on repeat. More importantly, it would queue videos based on how long viewers had watched them instead of how many clicks and views they had gotten.

YouTube's goal was for users to "watch more and click less," meaning they didn't want viewers to have to click on a bunch of videos before finding what they wanted. The AI could match them better to content they liked so they could spend more time actually watching.

This Watch time switch transformed YouTube's viewership—people did stay on the website longer. Misleading "bait-and-switch" tactics used by some creators were no longer being rewarded by the AI, because viewers left quickly when the content didn't deliver what the title and thumbnail promised. Viewers did stay to watch videos that delivered what they said they would, and the AI kept track of these videos with longer view duration and recommended them more. Additionally, viewers stayed to watch what the AI recommended next because they were relevant to what they had already shown interest in.

In other words, viewers were taking this new AI bait: hook, line, and sinker. The new YouTube AI got visitors to stick around, and the YouTube folks were over the moon about it. They had been meticulously observing the data from the switch and waiting with collectively bated breath to see if it would work or flop. By May 2012, just a few short months after the new AI integration, the data showed that average watch time was *four times* what it had been the previous May. Collective sigh of relief.

The YouTube AI has changed over time to create a personalized feed based on customization. Its Homepage is no longer channel dominant but filled with a mix of videos directly chosen based on individual viewing patterns and behaviors. It now suggests, with uncanny accuracy, what a viewer *might* want to watch. This is a huge change from its surface recommendations. You're no longer dipping from the site (if you don't know what dipping is, ask a Gen Z kid) because the videos are just another version of the one you just watched—you're sticking around to click on the video that you've never seen before but are definitely drawn to. It's as if YouTube hired a tailor to come in and take your measurements so he could build you an outfit you didn't even know you wanted. Who doesn't love the feel of something that fits like a glove? And that also doesn't look exactly like every other outfit you own?

Diving Deep into the Deep Learning Machine

To explain further, let's rewind and reexamine the data. After the turn of the first decade in the twenty-first century, YouTube came face-to-face with some hard truths. First, their users were watching videos from a bunch of other platforms instead of coming to the site directly. YouTube viewership was up, but only because people were watching YouTube videos that had been shared to big platforms like Facebook and Twitter. This made it impossible for YouTube to gather data about their consumers and to retain and monetize them.

Another tough truth was that YouTube had different operating programs for different devices and applications, so they needed to collect the pieces and reboot an operating system in one place, directly from the source. Shockingly, at the time, YouTube didn't even have a dialed-in system for analyzing mobile usage, which was an embarrassing realization because a huge percentage of viewership was mobile. Its digitally ancient mobile development was painfully slow, and something needed to be done about it, stat.

Enter InnerTube in 2012: an interdepartmental program at YouTube HQ created to revamp algorithms and development from the top down. InnerTube was resetting the system and observing its reboot in one place to ensure everything fell into place correctly and *quickly.* It was imperative that implementations be made quickly and could be tested before applying across the board. If a new change didn't work, they needed to pull it promptly without it crippling the whole shebang. Then they would tweak and try again.

Another vital piece to the reboot was utilizing deep learning machines. Google's AI had undergone several phases of development and usage, and it was getting better and better. Google's deep learning AI was now capable of using gigantic neural networks that got really good at things like recommendation and search. Deep learning goes beyond basic machine learning in that it's built to mimic human neural networks. It makes nonlinear conclusions.

The input data for deep learning machines on YouTube came from the behavior of its users and monitored not only "positive" viewer behavior, like which videos they liked and kept watching, but also "negative" behavior, like which videos they skipped or even removed from their custom Homepage or "Up next" recommendations from YouTube. Monitoring both the positive and negative behavior of its users is vital to the algorithm's accuracy. This neural network has gotten so good that it can even predict what to do with new or unfamiliar videos based on current user behavior. Saying, "It has a mind of its own," is not much of a stretch. The AI actually doesn't observe the total Internet behavior of a user; it only watches what happens on YouTube. This matters because it's what maintains its pinpoint accuracy in recommendations.

How?

Let's say you went to google.com and typed "steakhouses in Los Angeles" in the search bar. Does that mean the next time you go to youtube.com you want it to recommend videos on how to grill a perfect steak? Or that you want to take a video tour of LA? Probably not. But if you search, "How to grill the perfect rare steak," directly on YouTube's search bar and click on the first recommended video, the suggested videos that pop up next might be, "World's strongest man—full day of eating," then, "How to clean a cast iron skillet." These secondary videos don't have anything to do with steak, but do you see how that viewer would be a likely candidate to continue clicking? That's a deep learning machine that knows what it's doing. And YouTube and its ecosystem are direct benefactors, because when viewers watch more, everyone makes more money and gets more brand exposure.

A Machine at Work . . . and It's Working

YouTube recommends *hundreds of millions* of videos to users every single day, in dozens of different languages, in every corner of the world. Their suggestions account for 75% of the time people spend on the site.

In 2012, daily watch time averaged out at about a hundred million hours. In 2019, that average sits at a mind-blowing one billion hours a day. One billion hours of video content being collectively consumed by viewers on one website every single day! Over this seven-year span and thousands if not tens of thousands of tweaks and triggers, the deep learning AI has gotten *really* good at recommending videos to keep viewers watching longer. It has become an expert digital gardener who knows which product to harvest for each customer based on the videos they've been "feeding" on. You can be a YouTube master gardener, too, when you arm yourself with the right tools. Just hang on to your shovel, because we are still breaking ground.

4 The Algorithm Breakdown

You just learned a lot about the history of the systems that have run YouTube since its inception, and you know that those systems have become quite good at what they do. But what does that mean literally? When you go to the website, what do the systems look like as you navigate? To really grasp these foundational concepts, let's clarify what is actually happening when a site visitor shows up.

As soon as visitors arrive at youtube.com, they are being followed. It's like when you were a kid and went to your friend's house to play and their pesky kid brother just wouldn't leave you alone, but think of it this way: instead of being pesky, the brother quietly observes your behavior and accommodates your every whim. You want a snack, so he runs to the kitchen and returns with an apple. You say, "no thanks." So he takes the apple back and returns with a bag of Cheetos. You eat the Cheetos. Then you have a conversation about Han Solo, so he runs to the living room and plays *The Empire Strikes Back* for you. The next time you go to their house, as soon as you walk through the door he hands you a cookie and turns on *Return of the Jedi*. His prediction about what you might want to eat or watch is based on the last time you came over, and it's probably spot on. Oh, and also, you're probably going to want to go to their house more often with this kind of treatment. They know what you like. (Unless he recommends *The Last Jedi* or *Solo*, in which case you'll just go to the Zuckerbergs' next time because those movies stink.)

Let's say that in place of Cheetos, you wanted carrot sticks, and in place of *Star Wars*, you watched *The Office* reruns. The next time you showed up, li'l bro would offer broccoli and *Parks and Recreation*. The concept works no matter your preferences.

These examples help explain YouTube's goals:

- Predict what the viewer will watch.
- Maximize the viewer's long-term engagement and satisfaction.

How they do it is broken into two parts: Gathering and Using Data, and Algorithms with an "S."

Part 1: Gathering and Using Data

YouTube collects 80 billion data points from user behavior every single day. They gather data in two key areas in order to achieve the goals of the AI. The first area it observes is user behavior via metadata. It determines things about a video based on the behavior of the person whose eyes are on the screen and whose fingers are doing the clicking. "Satisfaction signals" train the AI what to suggest or not. There is a very specific list of these signals:

- Which videos a user watches
- Which videos they skip
- Time they spend watching
- Likes and dislikes
- "Not interested" feedback
- Surveys after watching a video

- Whether they come back to rewatch or finish something unwatched

- If they save and come back to watch later

All of these signals feed the Satisfaction Feedback Loop. This loop is created based on the feedback the algorithm is getting from your specific behavior. It "loops" the types of videos you like through its suggestions. This is how it personalizes each user's experience.

Gathering Metadata

To really get down to the details, here's an explanation for exactly how the AI gathers data. Observing metadata starts with the thumbnail. The YouTube AI uses the advanced technology of Google's suite of AI products. It operates a program called Cloud Vision (CV). CV uses optical character recognition (OCR) and image recognition to determine lots of things about a video based on what it finds in the thumbnail. It takes points from each image in the thumbnail and, using billions of data points already in the system, recognizes those images, and feeds that information back into the algorithm. For example, a thumbnail including a close-up of world-renowned physicist Stephen Hawking's face is recognized as such in CV, so that video can be "grouped" in the suggested feed along with every other video on YouTube that has been tagged under the Stephen Hawking topic. **This is how your videos get discovered and watched.**

In addition, CV utilizes a "safety" tool that determines, based on the data it has gathered from the images in your thumbnail, if your video is safe for all audiences to watch, or if it has adult themes, violence, or other questionable content, and it gives a "confidence" score of that determination. This score also reflects how accurately the content matches what the thumbnail shows. This means that you can create a thumbnail, plug it into Cloud Vision, and know before you finalize your video upload how the thumbnail will likely be

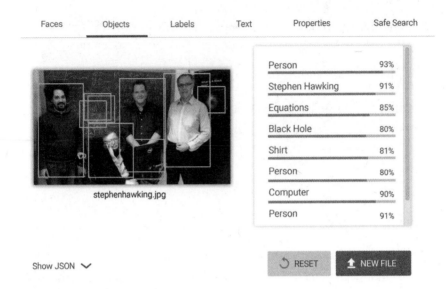

Figure 4.1 Thumbnail with data points

rated in the system. Using Cloud Vision can help catch something that might, for whatever reason, be flagged as inappropriate on any data point, and therefore can give creators the opportunity to fix it even before it is live. This has cut down on demonetization and other issues creators have had in the past. It can be a very valuable tool to help you stay one step ahead of the problems. CV is not an exact replica of YouTube's safety measures, but it is close enough that creators can get a good idea of how the content will be determined by YouTube. CV might tolerate something YouTube will not, but it is still a sufficient prelaunch tool to utilize.

Video Intelligence

Once the thumbnail has been checked, the AI goes through every single frame of the video and creates shot lists and labels based on what it sees in the content of the video itself. For example, if you do a video in a parking lot, the AI detects the store front, people, flowers, brands, and more, so it can log that info for recommendations and run it through the same safety routine that it uses to check thumb-nail images. Be aware of what is in the frame in every scene of every video you create! It will be detected by the AI and sorted accordingly, because the AI is validating the thumbnail. The AI cuts through the "noise" of every single thing in every frame and determines what is most important according to that video and its metadata.

Closed Captioning

The AI does the same thing with the language of the video. YouTube has an auto-caption feature now, and the AI reads through the words of the caption to gather data as well. So basically going through the video frames using shot lists is like looking at what is visually being said, while listening to the audio provides even more feedback via what is actually being verbalized. *Everything* goes into the system.

Natural Language

The AI is also listening for actual sentence structure and breaking it down into a sentence diagram. This extracts the meaning of what is being said. It can differentiate language so it can group it categori-cally, but not just on the surface. For example, two different creators might both talk about Stephen Hawking in their videos, but one video might be biographical or scientific while the other might be humorous or entertaining. Even though both videos are talking about the same person, they are categorically different enough that the AI would categorize them differently and group them with different recommended content because of the language being used.

Video Title and Description

As you should expect, the algorithm also looks at the video's title and description to supplement what it has already learned from the thumbnail, frame by frame, and the language. But it only tracks this as long as it needs to before it will use viewer data that comes in. The AI "knows" that people can be deceptive with metadata, but they can't lie about what's actually in the content. Don't slap a title and description on your video haphazardly just to get it finished and uploaded. The verbiage matters, so choose your words wisely. Most creators don't leverage the video description to its fullest potential. It's another data point the AI looks at to help with search ranking and discovery.

Part 2: Algorithms with an "S"

Did you know that YouTube has more than one algorithm? The AI uses multiple systems, and each has its own objective and goal. The surface features viewers see are:

- Browse Features: Homepage and Subscription
- Suggested
- Trending
- Notification
- Search

Each of these features runs separate algorithms trying to be optimized for a higher hit rate, and they all feed into the YouTube AI. They have separate hit rates to determine what actually works for users in each particular system. Hit rate means how often viewers are able to find what they actually want to watch. Have you ever heard of a fisherman getting a "hit?" It's when a fish takes the bait. Imagine that you are the fisherman who has tossed his video into the water. Potential viewers are the fish swimming by your "bait." Maybe 10 fish take a look at the bait and swim on by because it's not the brand of bait that they like. But along comes a fish who says, "That looks good," and he bites. Say you toss this line 10 times, and while 100 fish swam past, 10 took the bait. There's your hit rate. This hit rate is so important to each system in the AI. The algorithms are very sensitive to user behavior and the metadata on each traffic source so that they know how to change to increase the hit rate.

Additionally, YouTube is constantly running experiments—several thousand a year—and they implement about 1 in 10 changes as they go, so this translates to hundreds of changes being implemented annually. These changes help the system get smarter, and smarter means better at feeding viewers what they will watch.

Browse: Homepage

YouTube's Homepage has changed over time. Users no longer have to type a query in Search or to put in the work to navigate. The Homepage used to be where users saw only video recommendations of channels they had subscribed to. Now the Homepage has a personalized recommendation feed based on that user's history over time.

As long as a user is logged in when using YouTube, the algorithms can keep track of what videos that particular user has watched in the past so it can make better suggestions for videos that user is likely to watch, even if they haven't watched those videos in the past. This seems counterintuitive. It seems like the algorithm could be more successful suggesting videos it knows a user has already watched and liked, but actually the opposite is true. Keeping suggestions fresh actually gets users to stick around on the platform longer because they aren't getting bored with the same old stuff.

How YouTube does this is by breaking down the Homepage into two categories: familiar and discovered. It shows users familiar content from places the viewer has gone to before. These suggestions could include trending or recent videos from a channel a viewer has already watched. The discovery side includes videos or channels that users with similar viewing patterns have watched and liked. YouTube has found that these combined strategies are keeping viewers better engaged. If you want to get on the Homepage—and this definitely should be your goal—learn the triggers that get videos there. Try to improve your click-through rate and audience retention, because these will help you get pushed out to a more general audience.

Browse: Subscription

This one doesn't need a lot of explanation. The Subscription section pulls content from channels you've already subscribed to. It will suggest new videos from your subscribed channels, especially new videos with similar content to what you've consumed before. For example, you watched a prank video or two from a channel you have subscribed to, so the AI pulls that channel's newest prank into your Subscription feed.

Suggested

Another place besides the Homepage creators should be focusing on is Suggested feed, including the "Up Next" video. These are the suggestions below (on mobile) or to the right (on desktop) of the video that is being watched. This is a powerful place to be! Viewers stick around when this feature is working really well, and sticky viewers are YouTube's goal. So if you can use the triggers to get your video in the Suggested feed, you're exactly where you need to be.

What are the triggers? First, make sure you have created a strong relationship among the data in your own content. This means that if the metadata connects among your videos, the algorithm plugs your videos into the Up Next feed, and your video's likelihood of being watched skyrockets. Metadata includes title, keywords, description, and the content itself. We talk in depth about metadata in Part III of the book. The viewers' behaviors trigger which videos get put in the

Suggested feed as well, so when your content keeps viewers watching instead of bouncing, your content is more likely to be recommended here.

Other things the AI looks for in this feature include the "rabbit hole" type and the "watch something else" type. Rabbit hole explains itself pretty well. It's the type of videos that are similar in one way or another that keeps the viewer following a specific path. These include:

- Videos from the same channel
- Videos and channels that are similar to the one playing
- Videos that other people watched after watching the current video

The watch something else type (also self-explanatory) exists because viewers eventually tire of watching videos with similarities, and they need something entirely different if they are going to stick around. This isn't a random selection; it's still a personalized recommendation based on their past behavior. This recommendation comes when the AI has a stored history of what that viewer has watched over time.

Trending

I like to call this the geo-specific water cooler. A common misconception is that "trending" is synonymous with "popular," but it's not. Trending topics are broad topics that people are talking about right now all across the Internet. It is what's currently happening in the news, on social media, websites, blogs, and elsewhere. By "geo-specific" I mean that even the Internet has geographical regions, both in location and in authority. Websites with more authority than others will have topics that trend better because YouTube knows more eyes will be on them, so it pulls those topics into their own Trending section. They realize it's what people are seeing elsewhere so they're more likely to click on that topic when they are on YouTube as well. By location, groups that live in the same area often are interested in the same things. For example, something that would trend well in LA likely would fall flat in the Midwest.

Notification

When someone subscribes to your channel they can be notified by YouTube when you upload a new video, but only if that subscriber has also clicked on the bell button. The notification comes through on the subscriber's YouTube app or via email.

Search

Another straightforward feature is the Search. Users type in a keyword or phrase as a query to find what they want to watch. The Search feature displays videos related to that query. The algorithm narrows down the results based on the metadata and the video generated by the creator, and it also looks closely at past data from people searching similar queries, and how they responded to those videos. A lot of people think they just need to do SEO to make their video go to the top, but there is a "freshness" feature that will pull new videos into these results. Take a look at what's trending and create content with the right connecting metadata. Know especially what's trending in your niche. If your video performs well, the algorithm keeps it in the Search results, but if it doesn't, it drops it.

It's Not YouTube's Fault

You have just learned a lot about YouTube's history and its inner workings . . . congratulations! You've done some good digging. But do you find yourself musing, "Okay, great, but what does all of this mean for my *content*?"

I often come across creators with good intentions and lofty goals, but they aren't seeing the results they want to see, and they blame YouTube for it. If I can get anything from these first chapters to stick in your head it is this: *Don't blame YouTube, blame your content.* I know this sounds harsh—like I just called your baby ugly—but try to take a step back and look at it objectively. Maybe your baby really is ugly. If you can swallow the hard pill that maybe your content is to blame instead of the Big Bad Algorithm, you'll be ready to learn the YouTube Formula. If you won't consider that you might be doing something wrong, then nothing I say will help you, and you might as well close the book.

Now that you understand the inner workings of the algorithm a little better, you can move through the rest of this book ready to implement systems and strategies that work with the algorithm instead of being subject to intimidation or ignorance of it. YouTube's objective is simple: engagement and satisfaction with viewers. At the end of the day, it's all about creating good content. The purpose of this book is to teach you the formula to align yourself with YouTube's goals, and to be able to analyze how your content performs so you can adjust to achieve viewer satisfaction. You'll be ready to arm yourself with the tools you need to dig in and plant the right seeds in your content garden.

PART II The Opportunity

5 Why Most YouTube Channels Fail to Succeed

It's probably safe to assume you learned a lot about YouTube from Part I that you didn't think you came to this book for. Some of it may seem irrelevant to you at this point, but I promise it will all come together in the end. But let's be honest, you're here because you want to achieve massive success on the platform. Some creators define success as "x" amount of views or subscribers, but all YouTube success stories point to one thing: making money.

YouTube provides a place for people to share their passion with the world, and this is great, but do you know what is also great? When people get compensated for their passion. Making money is not a bad thing. *This is why we are here, folks.* You can spread your passion and make money doing it without being a sell-out. In fact, the more money you make, the more you can spread your message to the world.

Maybe you've been working hard toward achieving YouTube success for a long time, but your efforts remain fruitless. Remember how I said your baby might be ugly? With this thought in mind, I invite you to put down your defenses and be open to what I am about to tell you:

It's not YouTube; it's you.

I have helped hundreds if not thousands of YouTube creators see things about their channels that they simply could not see on their own. I have learned from many YouTube "failures" over the years in order to change the things that were not working. I use quotation marks because I do not actually consider them failures; I consider them an important part of the process. It's like Thomas Edison said, "I have not failed. I've just found 10,000 ways that won't work." "Failure" is a harsh teacher, but it is a teacher, nonetheless, and when you begin to accept what you might be doing wrong, you are ready to move forward. Some YouTubers don't analyze why a video underperformed or they don't know what to look at. So, what do they do? The same thing over and over again, hoping for a different result. This is called insanity, and it will lead to YouTubers burning out and either quitting or scaling back. I've seen this happen a lot. Making a living on YouTube isn't easy when you don't figure out why your content is underperforming so you don't know how to make small tweaks to course correct.

Most creators I work with have a handle on knowing what they are supposed to be doing, but they continue to put out content with the assumption that something will magically start working or suddenly attract an audience. This just isn't how YouTube works. Taking this approach leads to frustration and self-created hurdles that stop or slow you down from getting to your goal. Your first step to YouTube success might be to admit that your content is not working. Take a giant step back right now and ask yourself if you are actually happy with your content. Are you? Is it working? Do you know why it's not? One way to look at it is through the eyes of your potential viewer. Do you think they would watch your video, or even just half of your video?

As a company, YouTube is a prime example of learning from things that aren't working. Remember that it started as a dating website and changed based on "failure" feedback. They couldn't get women to post on their website even when they offered $20 a video. The dating website didn't work, so they changed the website to what its users wanted. If what you're doing hasn't been working, be willing to change it. Some creators who have been on YouTube a while often get trapped in old habits. If you fall within the "old YouTuber" category, reassess your ideology and methodology. Holding on to outdated processes and ideas could be what's been holding you back from seeing real growth. Just because something used to work for your content doesn't mean it will work forever. Don't be afraid to shed the "old" YouTube and look at how current successful creators are doing things.

The next step is accepting the fact that the data does not lie. Maybe your mom tells you that your videos are amazing, but have you considered that she might be the only one who thinks so? I can't help but think of those unfortunate American Idol hopefuls who walk into the room of judges with their mama's glowing compliments on their shoulders only to be escorted out of the room after an audience-cringing performance and a harsh critique from Simon Cowell.

The good news for you is that you can learn how to analyze data and create amazing content a lot easier than a bad singer can learn how to find perfect pitch. It's all about the data. You have to be willing to become a student of the data.

Become a Student of the Data

In order to become a student of the data, you must change the way you approach everything on YouTube. Stop watching for entertainment, and start watching to learn. Pretend you've enrolled at YouTube University and paid tuition to be there. This might surprise you, but a lot of creators don't even watch anything on YouTube; they only go there to upload their own content. I had a consultation with a creator who wanted advice to improve his videos. I watched two of his videos and asked him, "How often do you watch You-Tube?" He said, "Never. It's not what I like." This is like wanting to compete in the Tour de France because you have a bike, but you've never seen a race. And your bike is a Huffy with a banana seat. You don't have the right tools, and you haven't learned by watching others who do it well. Is this you? If you want to do YouTube well, you have to watch what other creators do on YouTube to see what works. If you never go to the Trending page, how will you see trends? How will you be able to see what's working and why? A YouTube student consumes everything on YouTube with a desire to understand how successful content works. This is achieved by looking for patterns.

Looking for patterns includes observing content, audience, and all the little details. To ensure we are on the same page, you need to understand that content is more than just the video itself. When I say content, I mean the video's title, thumbnail, description, hook, and, yes, the video. Start actively looking for patterns in each of these content breakdowns on channels that do it well, and I promise you will see them. There is a reason why they are getting the views . . . can you figure out why? Do the same with a successful video or channel's audience. I always dive deeper and ask questions, like: Who are they? Where are they coming from? What is their age and gender? What kinds of things are they saying in the comments? What other videos are they watching?

In addition to learning patterns, you also must learn the tools and systems YouTube gives you as a creator. There isn't a successful creator out there who doesn't know how to observe and interpret analytics. Stop being afraid of your analytics! The only way to become an analytics pro is . . . wait for it . . . by looking at your analytics. This is not rocket science. I realize this may seem easy for someone like me but that it might initially seem overwhelming for some creators. Again, all I ask is that you are willing to listen to what I will teach you and willing to do the work.

I was visiting my client Jimmy Donaldson, aka "MrBeast," in North Carolina when another creator named Zachary Hsieh, aka "ZHC," came to give Jimmy a custom Tesla. We were getting sushi at the end of the day, having a conversation about our favorite topic: YouTube data. I watched a few of Zach's videos and asked him a few questions about his data. Then I made a bet that I could guess his channel's demographic. Jimmy said he would pay for dinner if I could be within 2%. The bet was accepted.

Well, I've been on YouTube a while, and I've seen a lot of data. I always look for patterns, and I've been doing it a long time, so I thought I could guess right. After learning just a couple of things about the channel, I made an educated guess that Zach's viewers were 46% US-based and skewed more male to female. He pulled out his analytics, and guess what? I was right on; it was exactly 46%. They were shocked at my accuracy, and I was stoked that I had won the bet, because Jimmy had to pay for my dinner. Nothing is better than winning *and* getting free sushi! Thanks, Jimmy.

Know your analytics. It's the only way to know what needs to change, and something always needs to change. Even the most successful channels keep adjusting to be better. It's how they got successful, and it doesn't stop once you've "arrived" at success. Your audience will change and evolve over time. If you don't watch this, the audience will stop watching your content if you're not course correcting, because your content will become stale. There are so many ways you can learn how to improve your content, but the best ways are always based on what you can observe in your analytics.

In order to create your best work and make the smartest decisions, you have to gain an understanding of your specific niche. How do you do this? Do your research on other successful creators in your niche. Ask why they are successful. Take note of what they have in common. Why do they create the video and do the edits like they do? Watch their most successful videos and take note of these patterns. See if and how they engage with their communities, both in the video comments and on the Community tab. Read the comments on their most popular videos and note which specific comments are the most engaged. See what is being said, and you'll understand the viewer so much better by reading what they have to say here. Look at all content—title structure, thumbnail and description patterns, and video composition—through the lens of analysis and application, and you'll be ready when it's time to create your own content.

The Comparison and Copycat Traps

A word of caution, however: don't get caught in the comparison trap. Yes, you need to do niche-specific research, but that doesn't mean that you sit and compare your channel to others. I can't tell you how many channel consultations I've had where the creator tells me their content is so much better than their competitor's content. They don't understand why a channel with lower-quality videos would be outperforming theirs. Stop doing this. Instead, focus on what you can learn. Ask yourself what that creator might be doing to get more Watch time and viewer engagement than you. Focusing on the thing that makes you mad or frustrated makes it hard to observe analytically. Take out the emotion and put on your thinking cap. Comparison is a slippery slope and will leave you feeling discouraged and immobile. Focus on yourself and ways you can improve. I often laugh when I have conversations with MrBeast and he is comparing an old video (in the last year) to new videos. About one old video, he said, "Why would anyone watch this?" thinking it was a bad video (and it had 50 million views). We could have done it so much better knowing what we know now. Then we discuss how we could have made videos better based on the data. This is the superpower. You can always learn so much when you are competing with yourself. Always be improving.

Further, beware of becoming a copycat creator. There is a difference between following a trend and flat-out stealing an idea. You can learn from others' methods, content, and themes, and utilize them without copying exactly what they say and do. I have seen too many clone videos from copycats who produce frame-by-frame replicas of other creators' videos—don't be this person. Take an idea you like and put your own spin on it. Make content unique to your audience even if it's similar to what others are doing. Casey Neistat and MrBeast copycats are a dime a dozen; be you.

Course Correction and Consistency

So many creators don't know how to course correct when things are going badly, or even when they're going well, for that matter. Maybe they get a burst of views and then plateau and dip back down. They don't know how to analyze what went right or wrong, and they don't know how to replicate good data. They just continue to upload content without learning how to tweak to maximize results. This is why it is so important for you to break free of the dreaded "analysis paralysis." Do not be afraid of analytics! Some creators have become so afraid of seeing negative results that they resort to not releasing any new content. This obviously does more harm than good and absolutely hinders creativity.

Another reason creators run into problems stems from a lack of consistency. Creators need to follow a consistent schedule with uploading at the right time on the right days. There is no recommended schedule for all channels everywhere. For example, a Monday, Wednesday, Friday at 8 a.m. schedule doesn't work for every channel. Each creator has to test upload timing with their specific content to see what works best with their own audience. When creators are inconsistent, it is difficult if not impossible to get metrics feedback. Figure out what content resonates with your audience and when, and stick to it. This is how you get metrics data that will show you how to change or improve.

Being this consistent doesn't matter as much if your traffic comes from Search. For YouTube recommendation traffic—which is where you want to be—it does matter that you upload and interact when your audience is actively on YouTube and ready to engage.

Take the gaming channel *Thinknoodles* as an example. With its hefty five million–plus subscribers, *Thinknoodles* was operating under the *Field of Dreams* directive, "If you build (or upload) it, they will come," uploading new videos unscheduled. A subscriber base in the millions provided enough viewers stopping by, but inconsistency crippled its potential for velocity.

The *Thinknoodles* creator asked me to look at his channel, and based on what I saw in the data, I strongly recommended an ideal upload time for his audience that didn't make sense to him. It was a gaming channel, so he had been uploading in the afternoon when gamers would be getting home from school and work. But I saw that his content performed best in the early morning, so that's what I recommended to him. It defied logic. He gave me some pushback, but in the end, when he followed my advice, that video was the highest performing video he had had in a really long time. Once he corrected his course and shifted his strategy, his channel exploded to 64 million views in 28 days. Keep in mind that it was only a small tweak that got him millions more views in less than a month. See Figures 5.1 and 5.2.

Getting the Right Feedback

Now that you've become a student of the data and realized some things you might be doing wrong, you can put your energy into getting feedback. It is so important to get the right feedback from the right places (read: not your mom). Some creators mistakenly assume that getting any feedback is good, but not all feedback is created equal. If you are getting feedback from a group who doesn't represent

Views are up! Your channel got 64,770,178 views in the last 28 days.

That's more than the 20,490,000 - 25,190,000 your channel usually gets

Figure 5.1 More than 64 million new views

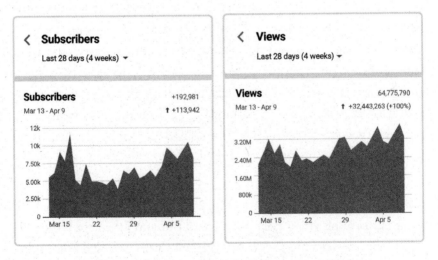

Figure 5.2 Increase in views and subscribers over 28 days

your ideal viewer then you could be doing more harm than good, because they aren't interested in your content the way that your ideal viewer is. So their behavior when they interact with your content is erratic, sending all the wrong messages to the AI and messing with your analytics. So if your audience is primarily Gen Z, you shouldn't be getting the bulk of your feedback from millennials and boomers. You're better off not getting any feedback at all rather than getting it from the wrong source.

Or maybe you're getting feedback from a group that just doesn't quite understand how YouTube and its AI work. For example, a small outdoors channel decided to make videos shooting bullets through objects to see if they were bulletproof. The creator's current content was based on walking around in the woods, so this seemed like a good related topic to his current videos. His accountability group liked the idea and helped him with video content ideas.

But there was a major issue with this topic. From where YouTube sat, gun violence was a fresh wound. They had had their own mass shooting recently at YouTube HQ, and there was a lot of turmoil around guns in the United States from repeated mass shootings and other things. What this creator and his accountability group didn't take into consideration was that you have to create content that You-Tube and its AI are "happy" to push and promote to viewers. Basically, you are a partner with YouTube; you have to pair your content with what they will be okay promoting. The successful content in this niche was from older channels who had already established their content with YouTube.

The thumbnail this creator put on his first video showed him pointing a gun at himself with a shocked look on his face. It was his worst performing video ever. He reached out and asked me why, so I helped him understand the reasons I listed earlier. He had gotten the wrong feedback. His group had meant well, but they didn't understand how YouTube sees and recommends content.

Another way to get poor feedback is from a group that is only going to pat you on the back and give you encouragement. There is a place for surrounding yourself with positive people, but when it comes to getting honest feedback for your content, Positive Polly isn't going to cut it. There are a lot of people who love to give feedback even when they have never uploaded a single video. Sometimes this person is your viewer. Trust the people who actually know what it takes to make a good video.

The Wrong Content

When it comes down to it, a lot of creators are putting out the wrong content for who they are. Sometimes "wrong" means a creator is doing something he or she is not passionate about. Don't waste your time doing something you don't care about just because you think it will work.

I have a friend, let's call her Sally, who has been on YouTube for 10 years, and who is probably reading this book. I have given Sally this advice for years. She started out as a daily vlogger, then the next month, she switched her content and became a sketch comedy channel. Then a few months later, she changed again and became a challenge channel. After that, she wanted to do serious interviews. *Stop doing this.* Figure out what you are passionate about, and quit changing with the trends. Spend more time figuring out what content you actually want to make.

Think about how the YouTube AI responds to a channel like this. If I'm confused as a viewer, what do you think is happening with the AI? It gets so confused about who her audience could be that it doesn't know who to recommend it to. There are no patterns to go from.

Sometimes "wrong" content means random. This is when creators aren't creating specific content for a niche; they upload whatever they want like they're throwing everything into Dropbox. Variety channels like this simply are not effective.

In addition, sometimes "wrong" means content that doesn't keep viewers watching. Content without retention doesn't convert. To keep viewers watching instead of dropping off, ask yourself why the content isn't resonating. Don't think about yourself, think about it from their perspective. Figure out where the drop-off is, figure out patterns where the content does resonate, then bring back that content into your creative process and make more content like it.

YouTube Superstitions

Some creators think there is a YouTube wizard behind a curtain somewhere pulling levers and calling all the shots, controlling content at his whim and pleasure. Of course, you have already learned that this is not true, because you have become an analytical student of YouTube who knows that the data doesn't lie. But others continue to waste time and energy on things that don't matter. Let me tell you a story to expound.

There was a woman who cut both ends off a pot roast before putting it into her roasting pan. She did this for years. One day, her husband asked her why she did it, and she said it was the way her mother had always done it. So the husband went to his mother-in-law and asked her the same question. She replied, "Because my roasting pan was too small to fit a whole pot roast." The wife's roasting pan was big enough to fit an entire pot roast all along, and she had been doing the task unnecessarily for years (not to mention she was wasting perfectly good meat).

Similarly, creators expend a lot of wasted energy when they don't know there is a better way. For example, there was a time on YouTube when there was a glitch in the scheduling feature, so channel subscribers weren't getting notified about new scheduled uploads. Creators started uploading manually because they thought it was the only way to get around the problem. The glitch was fixed, but a lot of creators continued to think it was broken. They believed this for years and were slaves to the clock. No matter what it interfered with—sleep, day job, family obligations—they stuck to manual uploading. YouTube's fixed scheduling tool would have saved them so much frustration, but they just didn't use it. Imagine how much of their energy and time could have been spent elsewhere.

If you are cutting off both ends of your pot roast, so to speak, know why you're doing it, and if you don't like it, explore your options. Maybe there is already a solution, but you just don't know about it yet.

Burnout

A common condition I see among YouTube success hopefuls is creator burnout. People get excited about content creation, but the harder they work and the more they grind, nothing seems to work and they get no results. So they work harder, uploading more and more content because they think it's a quantity problem. Working like this is a great recipe for disaster and eventual burnout. It feels like there is no light at the end of a very long tunnel. It is incredibly rewarding for me to work with burnout "victims" and provide them with tools that help them feel hopeful again, and, ideally, who finally start to see success. Other creators try to do too much and end up jeopardizing their mental health. If either of these conditions strikes a chord with you, you have come to the right place. Stick around; I am here to help you.

Don't feel like you have to upload three times a week to be successful. My friend and YouTube creator Mark Rober is an engineer and inventor, and he uploads a new video only once a month. He still gets tens of millions of views per video because his content is really good. His videos take a ton of planning and execution, but he is having fun every step of the way. He did a video about creating the perfect squirrel-proof bird feeder that required a lot of time, but it's fantastic content, and it got millions of views very quickly. He has a video about skinning a watermelon that has more than 100 million views. Another video took a lot of engineering and troubleshooting, but he created a glitter bomb for porch package thieves, and the video has more than 80 million views. Go study Mark's channel and see why you don't have to kill yourself with a heavy upload schedule to be successful on YouTube.

Realistically, yes, Mark would get more views if he uploaded more often, but his channel is doing great, he's making great money, and he isn't burning out or struggling with mental health issues. I know several daily vloggers who uploaded day after day for years, and the load eventually became too much to carry, so they quit. Modify your schedule if it doesn't feel sustainable.

Remember that this is a business, and there are ways to lighten your load. You don't have to do it all alone. You can and should build a team. For your sanity, build a team. When I was starting my own business, my dad told me that one of the quickest ways to success was by surrounding yourself with people who would share your goal and help you protect your sanity. Then he taught me what I now call, "Work Week Analytics." I document everything I do in the day, everything from checking my email to posting on social media to running errands, and everything in between. At the end of the week, I pull out two different colored highlighter markers. With one color, I highlight everything I hate doing, and with the second color, I highlight everything that wasted my time. Then I hire someone else or several someone elses to do all the highlighted tasks. This will help you stay focused on what you love and what matters, and save you from burnout.

Focus

Don't let your focus be your downfall. Your hang-up might be that you have been focused on the wrong things all along. You think you need a better camera or other equipment (you don't); you comment on tons of videos for channel exposure (you shouldn't); you blame YouTube or someone or something for your lack of exposure and success (it's not them; it's you, remember?). If your focus has been misplaced, own it and move forward.

You are ready to set clear goals to guide your efforts in the right direction. When you do, it helps solidify laser focus on your purpose, no second guessing. It allows you to focus on the things you will do every day to work toward your imminent success. We are not reinventing anything here: this is a proven formula that works . . . if you are willing to be a YouTube student, analyze your video's performance, and adjust to make better content that will resonate with an audience.

6 Make Money Partnering with YouTube

So you're ready to make money on YouTube . . . wait, you can make money on YouTube? How do you make money on YouTube? And you do this full time? How?

These are actual questions I get all the time when I tell people what I do for a living. The general population doesn't know and doesn't get it. The first and easy answer is that you can start making money by becoming a member of the YouTube Partner Program (YPP). As you'll see, this is just the first drop in the bucket, but we'll get to the rest of it in the next few chapters.

You remember from learning about the YouTube ecosystem in Chapter 2 that this program began in 2007 and was like the great California gold rush. Advertisement revenue sharing had never been done before Google's AdSense on google.com and then youtube.com, so this was a huge, brand-new opportunity. Creators could monetize their content if it performed well enough, so they did all they could to achieve this. In other words, YPP raised the bar on quality content and reliable, perpetual quantity.

I already told you that YouTube finally shared their annual earnings, claiming $15.15 billion revenue in 2019, but all of that money didn't go directly to YouTube. A huge chunk of it went to creators in the Partner Program. The program started out with a 45/55 split, which means YouTube took 45% of the revenue from ads running alongside content, while the creator of that content took 55%. The percentages are a little more complicated now because many factors are taken into consideration, but it's still a good return for creators. Monetization has been the number one reason for YouTube's success. It also has helped people look at YouTube as a viable career path or business.

Let me say that again: sharing revenue with the everyday people putting content on their website is what created exponential growth for *YouTube*. Yes, this changed the game for creators, but think how it changed everything for YouTube, too. It was a genius move. YouTube has so much money coming in because they began giving so much away. And because they were incentivizing creators, creators made better content that led to more visibility and views. When the content is better, people want to watch it more. The Partner Program is a brilliant win-win-win. Yes, that's three wins: for the YouTube ecosystem to thrive, everybody needs to be winning in the revenue triangle, including the creator, the advertiser, and the platform.

Today, there is an endless rolling lineup of creators wanting to monetize, but in its earlier days, YouTube actually sought out creators to join the program. Shaun and Mindy McKnight were among these early YPP recruits. They had been running a popular hairstyle blog and had uploaded a how-to video to YouTube to embed on their website. Shaun and Mindy hadn't even been paying attention to what was happening on their YouTube channel *Cute Girls Hairstyles*, but it had gotten enough traffic that YouTube reached out asking if they wanted to join the Partner Program. Shaun and Mindy made some fun "bonus" money in the beginning of their monetization, but they had no idea what lay ahead for them because of ad revenue. Eventually, Shaun would quit his job, they would be the first YouTube family with millions of subscribers, and their kids would have their own channels with millions of subs, too. More on this in the next chapter.

Joining the Partner Program

Now that you know you want to make money on YouTube—even if it's just because I'm telling you that you do—let's talk about how you're going to do it. To get monetized and stay monetized, you must follow YPP's rules, no ifs, ands, or buts. If you break the rules, you can get demonetized, lose your Partner status, or even have your entire channel deleted. This is serious; I have several people a day who reach out to me asking for help because YouTube demonetized or deleted their channel. YouTube doesn't mess around with this program. It may be tedious to read through all the paperwork, but you need to become familiar with the Community Guidelines, Terms of Service, copyright rules, and AdSense program policies if you want to stay in good standing. You will likely find anything you need to know by going to YouTube's support page and navigating from there.

But first, to qualify. It used to be that you could monetize with just one video, but YouTube quit doing this because it brought in a lot of "junk" content monetizing video-by-video. They wanted to encourage videos that people would actually watch and channels they would actually subscribe to.

You need 1,000 subscribers and 4,000 hours of Watch time in the past 12 months to be able to apply for the Partner Program. You also have to have a linked AdSense account. Once you reach these requirements, go to Monetization in your menu and sign the terms, and you're all set. You will automatically be put in line to be reviewed for monetization approval. Approval used to be automatic, but now the review team can and will reject a channel if they don't think the content follows YouTube's guidelines. The usual turnaround time is a few weeks to a month, but of course, there are always reasons it might take longer. This is the part where you get to practice being patient, but it's a great time to double-check that your content and your channel follow YouTube's guidelines.

It takes a little while because real humans do the assessments and approvals, not machines, so allow some space for them to do their jobs before you go pounding on the door for answers. If you've crossed your t's and dotted your i's, it should simply be a matter of *when* not *if* you'll receive your YPP welcome email.

YouTube will continue to monitor your channel after YPP approval to ensure your content continues to fall in line with policy. No creator is entitled to the program just because they fit the numbers game. Remember that it's a partnership, and that you have to hold up your end of the bargain to maintain the perks of the program.

The review team will check your channel's main theme, most viewed and newest videos, titles, thumbnails, and descriptions to verify compliance. But that doesn't mean you can "get away with" letting other things slide, because really they can check anything, and if they find something off-policy, the video or even your channel can be demonetized. Just follow the rules.

The Money Logistics

Let me explain how the YPP money works: advertisers pay YouTube to show their ads on the site as display ads, search result ads, and as preroll, midroll, and postroll ads. When a viewer sees an ad on a video, the advertiser has been charged money that was taken in by YouTube and split with the creator whose video the ad played on. How much the advertiser paid is called the CPM, cost per mille, or per thousand impressions.

As a creator, you can go to your revenue metrics and see how much money you are making from these ads playing on your videos by looking at the RPM, or revenue per mille/thousand, metric. According to YouTube, they pull in your total revenue from ads, Channel Memberships, Super Chat and Super Stickers, and YouTube Premium and multiply by 1,000, then they divide that by your total views in the same time period. This used to be an "estimated" metric, but YouTube announced the new RPM metric in July 2020 to show transparency and reassurance to creators as numbers declined from Covid-19. I'm glad to report that the RPM metric is no longer estimated, and that you don't have to do the calculating yourself. It was always an awkward "estimated" metric, so it's awesome that YouTube fixed it. Thanks, YouTube.

It's helpful to know how much advertisers are spending (CPM) and how much you can make (RPM). Keep track of these numbers when you log in to YouTube, and you'll find yourself thinking about your content and strategies in new and exciting ways, focusing on the right data points that bring in money.

As the creator, you don't get to choose which advertisers run ads with your content; in fact, it's the other way around. Advertisers choose where their ads run, and they get what they pay for. Which means the more specific they get in their targeting, the more they have to shell out. For example, paying for an ad to be shown to "females" costs a lot less per mille for an advertiser than choosing "Asian females age 35–45, zip code 90210 in southern California." They can get crazy specific here. They can specify their target as cat lovers, or within a household income range.

Additionally, just because you've been accepted into the Partner Program doesn't mean you've earned the right to have great ads run with your content. Think about it: if your content isn't being watched by any of the demographic specifications that advertisers choose, the ads won't show up with your content because it doesn't meet the specifications. So in theory, you could be a YPP creator with no money coming in. What can you do about it? Know your audience, then figure out how to appeal to pop culture, too, which expands your reach to a broader demographic. We talk more about this awesome audience blend in Chapter 11.

Also, the CPM can fluctuate; it goes up as more people advertise. Often this surge in ads happens around holidays and big events. Think of all the ads you see during the Christmas season. When there are more ads, it costs the advertiser more money to run them at that time.

Additional YPP Revenue Opportunities

What else does the Partner Program offer? In addition to ad revenue, YPP provides other ways to monetize, including channel memberships, live stream Super Chat and Super Stickers, a merchandise shelf, and a cut of the YouTube Premium subscribers' fee when they watch your content. Each has its own set of requirements—some have subscriber count, age, and/or regional limitations—so find out what those are when you want to implement them. You will also have access to tools like the creator support team and Copyright Match.

Channel Memberships

Channel viewers can pay a subscription fee to join a channel membership, which supports the channel and gives the viewer special perks. Memberships can offer members-only content like early access, behind the scenes, and exclusive videos. They also can include things like chat badges or custom emojis so members can stand out in comments and chats. Some viewers pay for a membership simply to support their favorite channel(s) and belong to their exclusive community.

To apply for the channel membership program, you must be in good standing with YouTube. Gaming channels have to have

1,000 subscribers, but all other channels must have 30,000. You also need to be a YouTube Partner. The membership program is not available everywhere in the world, so check if it's available in your area. Memberships cannot offer some things, including one-on-one meetings, downloaded content, and marketing to children. Revenue is split 70/30 with YouTube, so you get to keep 70% of the membership fee while YouTube takes 30%.

Some creators make a lot of money with their channel membership programs. According to a 2018 *Variety* article ("YouTube Creators Getting New Options for Paid Memberships, Merch Sales, Video Premieres" by Todd Spangler), a musician made 50% of his YouTube revenue from channel memberships, and he planned to fund a world tour with this extra income. A couple of travel vloggers signed up members fast with the promise of an exclusive miniseries, gaining 1,000 members in six days across dozens of countries. Another creator more than tripled his comedy channel revenue after implementing the program.

Live Stream, Super Chat, and Super Stickers

Super Chat and Super Stickers are a fun addition to live streams and another way you can diversify your YouTube income. They are basically a "tip" from viewers. A Super Chat shows up as a pinned comment at the top of a comment feed during live streams. It is highlighted by color and duration depending on the amount tipped. This encourages the viewer to spend more money. And Super Chats encourage other viewers to follow suit, often creating a domino effect. When you add Super Chat to your live stream, introduce it to your viewers. Many of them don't know about it, so give a quick intro before jumping in to your live stream, and then let them know when you get a Super Chat to recall their attention to the option.

YouTube gamer PrestonPlayz likes that Super Chat creates another avenue of interaction between dedicated viewers and content creators. He does a great job at acknowledging his live stream contributors while still being engaged in the purpose of the live stream. Always give a shout-out or thank you to your Super Chatters.

My friends at *The Ohana Adventure* channel did a live stream family game night every Monday for two years. When they implemented Super Chat, it ended up bringing in enough money so that they were able to go full time with their YouTube channel. Not only were their followers giving them money, but they had great ideas for future content creation that performed extremely well.

I've witnessed a live stream that pulled in more than $400 a minute with their Super Chat. This is such a simple integration that can make you a lot of money while being fun at the same time. You can implement automatic triggers to respond when donations come in from different tiers. Tier breakdowns encourage viewers to donate more if they want a bigger response triggered. For example, you can deck your set with automatic lights and a confetti cannon, and when you get a Super Chat, it triggers the light to flash, or the cannon to fire, or both, depending on the amount the viewer donated. Join IFTTT.com (If This, Then That), where you can apply the automatic YouTube Super Chat functions.

Super Stickers are animated graphics viewers can buy to show up in the chat stream. It's a fun way to give your viewers donating options while helping them feel like they belong to your community. Stickers are purchased separately from Super Chats.

Merchandise

The YouTube merchandise shelf has brought in a ton of money for some creators. The shelf is featured underneath videos and offers more than just T-shirts. YouTube partnered with Teespring so creators could design and sell a variety of branded items like bags, pillows, blankets, stuffed animals, mugs, and more.

According to an article in *Promo Marketing* magazine ("YouTube's Teespring Merch Integration Is Paying Off Huge for Content Creators" by Brendan Menapace), merchandise click-throughs get 30% more traffic than other features like banner links, annotations, and description links. Mugs, stuffed animals, and T-shirts are always in demand for viewers wanting to support or showcase a brand. Utilize these timeless products to further diversify your revenue sources. An animated jumping spider named Lucas has more than 400 million views and sold $800,000 in plush Lucas replicas in the first 10 days from the channel's merch shelf! The plush doll sells for $20 a pop on Teespring, but other options include T-shirts, baby onesies, backpacks, and stickers that say, "Boop!" Lucas's catch phrase. Get yourself a merch shelf.

YouTube Premium

Viewers can pay a membership fee to join YouTube Premium for an ad-free platform experience. Premium also allows downloads and background play options. YouTube gives the majority of this revenue back to creators, which means that you'll get a percentage when Premium members watch your content. Even if viewers watch your content as a download or background, you'll get the same Watch time credit. I wish I had an affiliate link to plug in here to encourage everyone to sign up for YouTube Premium, because it is awesome. It is worth every penny to have an ad-free experience and the download and background play options!

Make Even More with Ad Revenue

Once you become a YouTube Partner, your ad revenue still has the potential to change and grow—not all ad revenue is the same. The

YouTube creator ratio is majority male, so there is greater competition for advertisers who want to work with female creators. Similarly, there is fierce competition and opportunity for reaching viewerships that focus on specific races, ages, religions, and a range of other demographics. So the exact same ad running on different content can cost different amounts based on who the creator is and who their audience is.

Google Preferred

For example, Chad Wild Clay and his wife and partner Vy Qwaint make content together and put the same types of videos on their channels. Chad's channel has millions more subscribers than Vy's, but Vy's channel makes more than double Chad's in ad revenue because advertisers pay more to be seen on a female minority channel. The "Chads" are a dime a dozen on YouTube, but the "Vys" are a hot commodity.

When Chad and Vy create a video they know is going to perform well, they upload it to Vy's channel because they will make a lot more money. Vy belongs to the exclusive Google Preferred Lineups program, which is reserved for the top 5% of all YouTube channels. Only top tier advertising clients run ads alongside Google Preferred creators, and vice versa; only top tier creators have access to premium ads and ad revenue, which could be three to five times more than the average CPMs. Further, there are even tiers within the Preferred program. You can almost count on two hands the creators in the uppermost earning tier—it's that exclusive. They are earning a lot more than other creators within the Preferred program.

I'm sure you're thinking, "Well, where do I sign up for that?" but creators can't apply to the Preferred Lineup program. YouTube collects signals from five categories to produce a "P-Score," which ranks content that meets Preferred standards. The five categories the

P-Score algorithm watches are (1) popularity (based on Watch time), (2) platform (content more frequently watched on large screens like TVs), (3) passion (engagement), (4) protection (content suitability), and (5) production (superior cinematic technique and camera work). This combination of signals wades through oceans of content to find the most engaging and brand-appropriate content, no matter what category that content falls under.

Brand-appropriate, suitable content is a big deal to YouTube, and with good reason: they are working to repair a shaky history with advertisers after some big fall-outs. Advertisers left YouTube en masse when their ads ran alongside unsuitable content and some even boycotted, so YouTube is going out of their way to prove to ad companies that they have made it safe and desirable to market on YouTube. Suitability is a big selling point to get advertisers to see the value in the Preferred program.

The P-Score algorithm is always at work and updating regularly, so even breakthrough talent can land on the list; there's not a Preferred program exclusion based on creator longevity. Members of the Preferred program see big growth in brand lift, ad recall, and purchase intent. Brand lift means an increase in audience awareness and perception of the brand. They'll more easily remember you and your ad, which is termed "ad recall." It also increases your audience's intent to purchase your product or service. In short, everything elevates from a good campaign.

Because you can't apply to become a Preferred creator, your best course of action to work toward it is to remember the five signals that feed the P-Score: popularity, platform, passion, protection, and production. Use these signals as you strategize content creation and goal planning.

And don't forget that advertisers spend more to choose specifics about who sees their ads. If they want to target viewers based on age, gender, ethnicity, location, interest, household income, or any further characteristic or demographic, they pay for every specificity. So even if you are not a Preferred Partner, you can still make more in ad revenue than another creator if your content pairs well with a precise niche or demographic. Do your homework to let this information work in your favor. Create content to hit those unique audiences if you want to get more of the money advertisers spend on CPMs.

YouTube Selects

Advertisers also have the option to pay for the YouTube Selects program. YT Selects collects a lineup of content that will match an advertiser's ideal customers and run their ad on that lineup. The "emerging lineup" feature additionally pulls in new/rising content and creators. YouTube guarantees brand safety and suitability controls as well, meaning brands can feel confident that their ads aren't running on inappropriate or questionable content.

YT Selects ensures that advertisers will get exposure on all devices and YouTube apps as well, so their ads will be seen where people are actually watching. They even offer brand lift measurements so advertisers can see how effective their campaigns are.

Case Study: Dan Markham @ What's Inside

Let's talk more in depth about the funnest part of YouTube with some awesome success stories. What's the funnest part of YouTube? Making money, of course! Take Dan Markham from the *What's Inside* channel, for example. He gets to hang out with his son and make a lot of money doing it. They cut stuff open for their videos, and they have made a lot of memories together over the six years they've been active on YouTube. Their multimillion subscriber channels provide ample income for the Markham family to live any lifestyle they want, but Dan remembers the early days of being a YouTube Partner and being excited about the tiny returns. His channel was making nickels and dimes in the beginning, so he remembers very distinctly the day when he made $4 in ad revenue . . . and felt excited about it.

"I thought maybe we could put some money away for college if we had a little income that was coming in while we were sleeping," Dan said. "So I started reading tons of articles to research how to do YouTube." He started putting up videos weekly, which is all he could commit to at the time with his day job and family life. In four months, he reached a thousand subscribers; in three more months, he had a hundred thousand, and just four months later, he had a million subscribers.

In those early days, Dan could see the potential for real growth with YPP AdSense, but he never dreamed of the domino opportunities that would be coming because of it. It opened up brand deals, sponsorships, and big business ideas. This would be exciting for someone who wasn't making a ton of money in their regular job, but Dan was. He had a very comfortable, stable job that paid well, and so did his wife. She was a top VP at Honeywell, and they were each making six figures (not to mention their benefits packages). Money wasn't a problem for the Markhams. So even when the YouTube channel took off and was making more than the day job, Dan felt uneasy about quitting something stable for something that felt like it might be a phase. Looking back, he can see that he could have gone full time on YouTube sooner, but he had to be comfortable with the idea that "stable" in a traditional sense can be crippling.

Case Study: Jackie @ JackieNerdECrafter

Sometimes losing is the best thing that could happen to you. This was true for Jackie "NerdECrafter," who entered a giveaway for a cute pin with a parrot on it. When she didn't win, she thought she would just buy the pin because she loved it so much, but then she saw the $80 price tag and had an even better idea: she could make her own.

Jackie had already tried her hand at a few YouTube videos but with no direction and no success. Mostly, it had been something to pass the time after she lost her job. Now she thought she could make a video of her cute parrot pin craft and teach other crafters how to make it.

At the time, female channels often fell under the beauty and makeup umbrella. Even the crafting channels gave off a perfection vibe. Jackie thought that in order to be successful, she had to do the same. It wasn't working for her. So she stopped doing it their way, and did it hers. As soon as she decided to stop trying to be perfect and just be herself, something interesting happened . . . her channel started to take off. Her authenticity rang true for an audience that identified with her down-to-earth "nerdy" side. She was a nerdy crafter who was into gaming and geek culture, and her sense of humor was very attractive to the right audience.

In Jackie's first year as a YouTube Partner, she did a total of about $100 in YPP AdSense. Her second year brought in $500. She was spending a lot of time creating and editing videos basically for free, and always in her spare time outside of her teaching job. Then something great happened in year three: her consistent, authentic content found her audience, and she made $23,000. Year four brought $34,000, and after that, it tripled. "I could see that there was a pattern," Jackie told me in an interview. "The more I was confident on camera, the more candid and goofy and genuine I was, the better my videos did." She took a sabbatical from teaching so she could treat her YouTube channel more like a business, while always keeping fun at the forefront.

Jackie's mom passed away when she was younger, and her dad didn't take care of the family, so Jackie had to step up and be the parent for her siblings. "It was pretty much the definition of a broken home growing up after my mom passed away because she was the glue, she was the joy of the house. My number one goal has always been to take care of my family. I don't want anything from anyone.

But in terms of the money aspect on YouTube, I want to give my siblings the life they should have had." Jackie has enjoyed taking her family on trips and seeing the joy on their faces when they get to experience new places. "I never would have been able to do that on my teaching salary. Honestly, teaching money at this point is pocket change."

Case Study: Shay Carl Butler @ Shaytards

Shay Carl Butler is an OG YouTuber who has been on the platform basically since the beginning. He was one of the first to make money from ad revenue. "I became a YouTube Partner in February 2008, and got my first YPP AdSense check April 24, 2008, in the amount of $367.40 for a month's worth of videos," said Shay on Forbes.com ("ShayCarl's Epic Journey to YouTube Stardom" by Michael Humphrey). "I *could not* believe that I was actually getting money for 'entertaining.' I knew that my family couldn't live on less than $400 bucks a month, but the wheels in my head were turning. I immersed myself into YouTube." Shay quit his granite countertop business, became a weekend DJ to make ends meet, and the rest is history.

Forbes called Shay one of the "most successful video entrepreneurs on YouTube." This is the guy who had been a door-to-door salesman, school bus driver, MLM guy, countertop installer, radio DJ, and college dropout because he could never find a parking spot. Now he is the guy who has sold a company with his partners to Disney for $500 million. All because of YouTube. *Side note: Shay would be the coolest bus driver in the history of all bus drivers!*

Case Study: Ryan Kaji @ Ryan's World

Forbes 2019 list of YouTube's highest paid creators carries a hefty collective sum (and the list isn't totally accurate, because some creators, like MrBeast, don't publicize their earnings). The numbers are mind-blowing, but the list is an eye-opener for those who don't comprehend the opportunity for growth here. Take YouTube child star Ryan Kaji, for example. His channel started as *Ryan Toys Review* and has since become *Ryan's World*. His "unboxing" videos shot him to YouTube stardom—in its first five years, Ryan's channels amassed more than 40 billion views collectively. Forbes listed Ryan's channel as #1 with $26 million in total revenue for 2019. But the ad revenue from the channel isn't their only stream of income. In fact, it's not even their greatest source of income. Let that fact settle into your brain for a second. Their ad revenue falls in the *tens of millions of dollars . . . and it's not their greatest source of income*. YouTube opened the door to a world of opportunities for Ryan and his family, and the same can be true for every other YouTube creator or business.

YPP AdSense Is Just the Beginning

You might roll your eyes and think, Sure, I bet it's nice to be the #1 channel on YouTube. Let me stop you right there. You don't have to be on a Forbes list like these creators to make a lot of money on YouTube. Becoming a YouTube Partner is a great start, and there is a lot of money to be made there, but you can do so much more than that. As you stay true to your passion and connect with your audience by using the YouTube Formula, you'll be ready for every opportunity that awaits.

YouTube CEO Susan Wojcicki reported at the end of 2019 that, "Compared to last year, the number of creators with a million or more subscribers has grown 65%, and creators earning five or six figures, annually, has increased more than 40%. YouTube as a platform for creators continues to thrive." She's not wrong. YouTube creators are continuing to capitalize on opportunities created on the platform.

YouTube provides a unique opportunity for creators to go beyond making a supplemental or full-time income to opening a whole world of possibilities. It is the opportunity to build a business and a brand. No other platform has the power or the means to share the revenue that YouTube has. And as you have learned here, ad revenue is just the beginning.

7 Use Your Influence to Generate Big Money

Jessica Hatch had been working full-time at her family's restaurant for 12 years and on YouTube for 7 years when her mom pulled her aside and asked her why she wasn't doing YouTube full time. This is an uncommon response from a parent or loved one. When a YouTuber wants to become a full-time creator, usually their family or friends respond more critically. The next day, Jessica trained her brother in the restaurant, and the day after that, she went all in on her YouTube channel *Gone to the Snow Dogs*, featuring life with her Siberian huskies. It was a scary transition for her to leave her routine and the stability of the day job—and she was making more than double as a YouTube Partner than what she had been making on shifts at the restaurant. Even now, after four years as a full-time YouTube creator experiencing multiple revenue streams, she still can't believe how much money there is to be made on YouTube, and how long it took her to realize the true opportunities from YouTube.

As you just read about in Chapter 6, AdSense from the YouTube Partner Program (YPP) is a great place to start making money on YouTube, but it's just the beginning. When Jessica first discovered revenue possibilities outside of YPP AdSense, she was floored.

She told me, "I had heard people talking about brand deals and sponsorships, and my first thought was, 'Whoa, I can do that. I can reach out to people.' So some of my very first brand deals were me literally picking up dog toys going, 'Who makes this? Can I talk to these people? Sure, I can find their email,' and I would just email people." What started out as product freebies eventually turned into big deals with companies like Sony Pictures, Disney, and others. We'll talk more about brand deals and sponsorships in a second.

Jessica's brand deal/ad revenue ratio has continued to climb to 70/30 in favor of brand deals—sometimes it dips to 50/50, but you can see the potential for huge opportunities beyond YPP AdSense in either scenario. Jessica has enjoyed incredible success with a very specific niche and audience, expanding her reach and being a part of opportunities she probably wouldn't have had otherwise . . . all because her mom was wise enough to give her the nudge she needed to live her passion. What a great mom. Most parents do the opposite, telling their kid to get a "real job," but most parents don't understand the opportunity for massive success on YouTube.

The *Slice n Rice* channel features the interracial relationship of Matt and Glory and is a great YouTube rags-to-riches story. Matt ("Slice") had been working a retail job he hated and was barely scraping by, and Glory ("Rice") had taken the big risk of quitting her job to focus on their YouTube channel. Matt felt a lot of pressure to find a way to provide, especially with their upcoming wedding. They had a couple of videos "pop," or get a lot of views fast, which generated a YPP AdSense income that happened to match exactly what they had planned for their wedding. It felt like fate or divine intervention. Matt gave his two-months notice at his job and committed to doing YouTube full-time with his new bride. This was a huge leap of faith and a big risk for them. From there, the channel and its ad revenue continued to grow, and Matt and Glory felt so grateful that they had put in the work for their passion.

The best part about this story (and most others) is what came after AdSense: more opportunities. One brand deal in particular made all the difference for *Slice n Rice*. Warner Bros. approached them to push an upcoming movie that featured a mixed-race couple. Matt remembers thinking that the amount of money Warner Bros. offered them must have been a mistake, so he didn't negotiate terms or rock any boats. They only had two to three hundred thousand subscribers at the time, and the deal felt like it didn't match their reach, that Warner Bros. must be overpaying. But they didn't know what Warner Bros. knew: how valuable a dedicated audience could be. After seeing the Warner Bros. movie's success and attributing it in part to their recommendation, Matt and Glory began to understand the power of influence—particularly their own power of influence. Their authentic creator-viewer relationship could make a lot of money for the right brands.

Matt and Glory have made exponentially more money on YouTube in a short amount of time than most people make from a yearly salary. "We kept going to our bank account and just staring at it," Matt said, "We didn't know what to do with it." And they are just getting started.

Why Not You?

YouTube success stories like these are not rare; I could give you thousands of examples, and we'd barely be scratching the surface. The point I'm trying to make is that *anyone can do it*. You don't need a lot of disposable income or fancy camera equipment or 60 hours a week to be successful on YouTube. You *do* need to learn about the platform, see how others have done it, recognize the endless opportunities, and follow the formula I'm giving you. Oh, and it doesn't hurt to have a little passion. Actually, if you don't have passion, chances are you will become one of the crash-and-burn statistics. I have seen many YouTube creators who burn out. I have seen many creators who rode the YPP AdSense wave but didn't diversify their income streams and lost everything. But I have also seen so many creators achieve great success and surpass every expectation they could have had with opportunities they never knew existed.

If you think there isn't a big enough audience for you, think again. There is a niche for everyone. A guy in my small town owns a 24/7 towing service company, and he started filming and editing videos on his phone that captured his towing rescues and escapades, and his channel has more than 26 million views in just over a year. Towing cars. No really . . . *towing*. He's now a student of mine, and I can't wait to show him the opportunities out there. Check out *Matt's Off-Road Recovery* and see that you don't even need the best equipment to be successful.

YouTube success can be reached by anyone, anywhere, with amateur equipment. If you are interested in something, chances are there are a lot of other people out there interested in the same thing. And when you find your people, you can have a valuable influence on them that can give you a lot of opportunities to grow.

Merchandising

Most people think merchandising on YouTube means you have a link for viewers to buy T-shirts and hoodies. Influencers like to call it "merch," but it can be a lot more than a T-shirt sale here and there. I did a merchandise strategy with MrBeast where he promised to autograph a batch of limited edition MrBeast shirts. The purpose was to raise money for a big giveaway video for his 40 millionth subscriber. He challenged the Internet to "do your worst," and it responded in kind. He sold 68,337 shirts, and it took him 12 exhausting days to sign them all. But now he had a huge budget to make an epic 40 millionth subscriber video.

Merchandising can be more than a "merch shelf." You can partner with a brand to make a product specific to your channel or your niche. Beauty influencer Jeffree Star has a huge following on YouTube. In 2019, he collaborated with YouTube celebrity Shane Dawson on a docuseries and an eyeshadow palette. The pair generated a ton of hype and exposure from their combined online presence, and the response was staggering. When the palette launched, it sold one million units in 30 minutes. At $52 a palette, they raked in $52 million. And they could have sold a lot more had they not sold out. The influx of traffic crashed Shopify for hours. Star had estimated sales in the millions, but I would guess that even he was blown away by the dollar amount that his influence translated to. Star owns Jeffree Star Cosmetics (JSC) and brought in roughly $210 million in 2019.

In Chapter 6, I told you about the world's highest-paid YouTube star, Ryan Kaji, earning $26 million in ad revenue in 2019. One tiny detail . . . he's eight years old. His parents have been running his channel, *Ryan's World*, since he was a preschooler. Five years later, the channel boasts more than 25 million subscribers. That translates into a lot of opportunities and money way beyond that mere $26 million ad revenue. Ryan has his own TV show with Nickelodeon called *Ryan's Mystery Playdate*. But here's the kicker: merchandising. Ryan's face is on a whole line of products at Walmart and Target—things like toys, DIY projects, toothbrushes, and even underwear. Retail sales for Ryan's products reached more than $150 million in 2019.

Brand Integrations and Influencer Marketing

Brands and businesses that understand the power of an influencer will spend the money to capitalize on that power. They know it works. People feel connected to something they care about, and they respond to authenticity. So when a creator shares an authentic passion with their followers, they can easily persuade them to do something, buy something, or support a cause.

Google created FameBit, which matches brands with creators. It's like a dating app that matches people, but its purpose is to help brands and creators to find each other. There are many agencies that do this. Ricky Ray Butler is a business partner of mine. We produce a TV series called *The Chosen*, and we also cohost a podcast called *Creative Disruption*. Ricky Ray knows influencer marketing as well as anyone out there. In fact, he started working with Facebook influencers when he was still in college, before digital media agencies were a thing (and, actually, before the term "influencer" was a thing, but we'll use the term retroactively here for clarity's sake). He quickly learned that you could get ROI with influencers connecting with brands—a new concept at that time.

He founded his own media agency, Plaid Social Labs, and started leveraging the influence of people who had a lot of friends or followers. Ricky Ray noticed that these young content creators were developing their own organic communities around raw or "unprofessional" content. This was at a time when you could have counted on one hand the number of people online who had more than a million followers. Influencing was brand new (this isn't even counting Instagram, which wouldn't be around for a few more years). Ricky Ray realized there was a whole world of valuable brand collaborations for someone to seize.

Traditional marketing agencies didn't track ads then; the bottom line was their only gauge. But Ricky Ray's agency collected data to track their clients' consumers. Their brand integrations were producing amazing results, and they had the data to back it up. Even so, when Ricky took his business model to board rooms, they scoffed at him. Here he was, a kid in his early twenties, trying to convince Big Business that young, unprofessional content creators could get people to buy stuff. The good old boys of marketing didn't know how to look at this new medium as a legitimate sales channel. One of Ricky Ray's first successful brand integrations was between Shay Carl Butler, a YouTube creator, and Orabrush, a new oral hygiene product. Shay Carl pushed the product to his loyal followers, and the collab pulled in millions of views for just a few thousand dollars. That kind of brand recognition and lift was extremely valuable, but the traditional marketers wouldn't hear of it.

So Ricky Ray kept doing his thing. Plaid Social Labs combined forces with Travis Chambers, social media supervisor at Crispin Porter + Bogusky, an ad agency hired to do a campaign for Turkish Airlines. You'll get the full story in Chapter 9, but in short, this ad was extremely innovative and unprecedented, both creatively and distributively. Travis and Ricky Ray's agencies connected the campaign with more than 800 influencers, who would push their followers to watch and share it. They used basically every top YouTuber at the time. The ad got more than 140 million views and three million social shares.

Ricky Ray did another campaign with an oral hygiene product called Steripod. Steripod knew they needed their ad to be inside the content, so they focused on YouTube influencer integrations specifically. Ricky Ray paired YouTube music star Lindsey Stirling and a couple dozen other influencers with Steripod, and in three months, the product had entirely sold out in all Bed, Bath, and Beyond stores across the United States and Canada. They had seen a 15% sales increase.

In 2015, Plaid Social Labs was acquired by Bill Gates's Branded Entertainment Network (BEN). BEN handles brand integrations and media influencers all over the world, and Ricky Ray became its new CEO. He reports to Bill Gates monthly. BEN's state-of-the-art technology in deep learning machines made it possible and even necessary to operate from an artificial intelligence angle. "We were forced to evolve into an AI company because the sheer amount of content and distribution these days makes it difficult to stay on top and relevant," Ricky Ray told me. In order for them to be at the cutting edge of the industry, they create a customized algorithm for every brand they work with. They customize each algorithm depending on the client's desired outcome: views, clicks, or actual sales. As a result, they have more performance and conversion data than anyone out there.

The agency worked with more than a hundred influencers to announce a new game called *Apex Legends*, which was released in early 2019. The game got 25 million players in week one. *Apex Legends* had crazy good downloads and conversions, but even more impressive was view predictions. BEN developed an AI that was able to predict 99.5% of the campaign's views with structured and unstructured data. When they launched the next campaign during a different season, they raised their prediction to 99.8% because they had more data to go off of. Being that precise had never happened before. If the data is there, it makes the job predictable; it's no longer guesswork. The agency is one of the best in the world at using data to harness the power of influencers in generating big results and big money.

The brand deals and integrations industry has taken a decade to develop and mature. Systems and processes for activating campaigns have to be completely different from the traditional way. Nowadays, brands have to be different, and they have to be data-driven. They have to be willing to learn what they don't know and what they've never done.

Branded Entertainment Network is the biggest company in the world that does influencer marketing and product placement, so they understand what's happening and what brands are doing that's working and not working. More than 80% of content on YouTube is from content creators—which means we are only seeing the tip of the iceberg when it comes to the potential of brand integrations.

Working with YouTubers can be a sales channel, but it can also be huge for brand awareness. Brands today can get as many views in a week as they do with a Super Bowl commercial, but it's a more effective way of marketing, because it's inside the content. Other ads across the Internet can be blocked by ad-blocking tools, usually browser extensions. On YouTube, ad inventory can be seen right with content that is being consumed by loyal, engaged, and trusting viewers. There is no reason why you shouldn't be capitalizing on these audiences.

On a personal level, Ricky Ray is passionate about his work because he thinks of it as a way to preserve and empower art. Throughout history, in times of peace and in times of turmoil, art has always been something that was of the utmost importance to preserve. In our digital age, content creators are our own version of the famous artists of the past. Ricky Ray feels like, "We are doing something important in getting their art out to the world in a way that can be consumed, appreciated, and archived. We want to empower our modern-day and future generations of artists and innovators." And brands have an opportunity to contribute to this movement. That's important work to make a career out of—the money is just the bonus.

Business Ownership

Beyond brand deals and sponsorships, there is an even greater opportunity to build your own business or brand. A lot of YouTube creators don't start a channel with this end in mind, but they should. This is where the biggest opportunity is for creators.

In Chapter 6, I introduced you to Shaun and Mindy McKnight, whose hairstyle-blog-turned-YouTube-channel changed their lives in ways they'd never dreamed of. Shaun had a comfortable job as international business director at Nature's Sunshine Products that he loved. Mindy was a stay-at-home mommy blogger trying to earn a little extra grocery and spending money. When YouTube asked them to join the ad sharing program, the McKnights thought it would be a fun way to bring in some side money. Soon, Shaun found himself weighing his options: should he stay at his steady-income job with insurance and a 401(k) but had basically zero opportunity to advance? Or should he quit his job to work with his wife, committing to a YouTube channel that made less money right now but had more opportunities for growth? He chose to quit his job. And it only took three months before their YouTube income grew beyond what his salary had been.

The McKnights had entered a YouTube channel competition called On the Rise. The monthly competition highlighted a few channels that were "on the rise," and the winner was chosen by popular vote. The McKnights' channel *CuteGirlsHairstyles* won, giving them a spot on YouTube's Homepage, and their channel was featured on Facebook and elsewhere. Their subscribers jumped 10% in one day. Their YPP AdSense paycheck that month had a 5,667% increase. That's not an exaggeration. Good thing Shaun decided to take that leap!

In 2013–2014, the McKnights' channel saw phenomenal growth for a channel at that time. YouTube had changed to the algorithm model that followed viewers, viewers' interests, and started recommending content to them, so the McKnights got more views because of this. In addition, they started seeing more ad revenue because of YouTube's switch to mobile views. YouTube also went international, so the McKnights leveraged this new global audience that was just discovering their content for the first time. YouTube was really figuring out their algorithm, and it was paying off for creators who already had the content the algorithm liked.

The McKnights' road to YouTube success was slow and steady. Their income started with just YPP AdSense but became 50/50 between AdSense and brand deals. AdSense and brand deals were great and could generate a lot of money, but they were sending traffic and sales to someone besides themselves. As great as the money was, it didn't compare to the true opportunity here: they needed to become their own brand deal. They started their own business and launched their own products, and if they could go back, they wouldn't have waited so long to do this. At the time, they worried about pushing products on their audience, thinking it would push them away instead of pushing them to buy. Looking back, they would skip the in-between stages of brand deals and licensing deals and go straight to ownership. "The audience likes you because you're passionate about your content," the McKnights said, "so they'll like your product because you're passionate about it, too." The authenticity goes straight through to the product on the shelf.

Mindy wanted to create a line of products that would be a one-stop shop for all different types of hair. And she wanted quality products that she would use herself. She formulated products and designed beautiful packaging for "Hairitage," and they pitched to Walmart. Forty-five minutes into their allotted hour, the Walmart people in the room said, in a nutshell, We've heard all of this before; what else you got? The McKnights then wowed them with their digital marketing knowledge and influencer reach. By the end of the meeting, Walmart wanted in (word on the street is they never commit like this in a first-pitch meeting), and they gave Mindy's products prime real estate on the shelves. With a retail giant like Walmart, there are default settings and safeties in place that control exactly what product brands can and cannot do. With Mindy, however, Walmart pulled out all the stops and gave her warehouse safety stock because her products could move quicker.

This was a home run for the McKnights, but it benefited Walmart, too. Mindy considers her Hairitage products in line with other brands you would find at stores like Target and Sephora, so she improved the quality of the products on Walmart's shelves . . . and the quality and quantity of the people coming into the store. A lot of Walmart shoppers go there for their grocery items, but now they would go for hair care products as well.

In addition, the traditional marketing Walmart execs operated old-school style, so they asked Mindy how to run digital marketing. Mindy taught them about "swipe ups" and affiliate links. For her products specifically, Mindy wanted micro influencers to push her products to their own dedicated followers. She wanted online chatter about the product, not a meaningless, nonproductive celebrity endorsement. The next step after micro influencers would be to get bigger influencers with bigger reach to tell their audiences. Mindy's meticulous care in aesthetic design resulted in products that were "Instagram worthy," meaning people would want to take a picture and post it. She knew it would be easier for influencers to have a bigger response if her products were pretty in addition to high quality.

Mindy's passion was hair products, so she created Hairitage, while her YouTube famous daughters Brooklyn and Bailey were passionate about mascara, so they created their own line of mascara called Lash Next Door. Mindy's advice to YouTube creators is to focus on your passion, not the money or the fame. If you're passionate and you push yourself to diversify and find new and exciting ways to grow, the money and the opportunities will follow.

Limitless Opportunities

Shaun and Mindy McKnight started on YouTube with small ad revenue. They went on to own six YouTube channels, collaborate with awesome creators and brands, and build multiple businesses. They have created products. They have negotiated huge deals. They have spoken across the world. And it all started with a blog on how to do hair. Examples of huge growth and big money can be found by the thousands on YouTube, in every niche and all across the world. Don't limit yourself. There is no reason why this couldn't be you.

8 The Real Power of Your Influence: Making a Difference

What do you want to be when you grow up?

Kids get asked this question generation after generation. The most common answers usually include job titles like astronaut, athlete, doctor, vet, and teacher. But kids these days are different, because they've grown up with the influence of the Internet, and the effect it's had on their answers is telling. In the United States, becoming a YouTuber is the #1 answer for kids these days. My son Bridger has wanted to be a YouTuber since he was in diapers.

And who can blame them? Successful YouTubers often showcase their glamorous lifestyles with fancy cars, huge houses, exotic vacations, and loads of endless fun. So many of them make it look like a life in endless paradise.

But not all successful YouTubers flaunt the lavish influencer life. Enter Jimmy Donaldson, "MrBeast," a kid who did make a successful career out of YouTube, but who kept his feet on the ground. When I found out about MrBeast, I followed him on YouTube and his socials. Then I noticed he followed me back. A few days later, he sent me a DM on Twitter saying he had watched some of my videos, that they had helped him understand the YouTube algorithm better, and that he would love to connect sometime to talk data. MrBeast at the time wasn't as big of a deal as he is now—he only had 4.7 million subscribers, but I could see that he would be creating big ripples on YouTube. So I messaged him back and said I'd love to connect.

Jimmy asked me where I was right at that moment, and I said I was working with a client in Dallas, Texas. He responded that he lived in North Carolina, but that he wanted to meet me immediately. He said, "I'm going to hop on a plane; I'll see you in six hours." Sure enough, he showed up six hours later, and we jumped right in, talking about YouTube and data for hours. Eventually, the conversation transitioned to the money side of YouTube. I asked him what type of car he drove, and he told me he drove a Buick. When I asked why he didn't drive a Lamborghini like a lot of comfortable YouTubers do, he said he wouldn't drive something he wouldn't also buy for his team. Instead of buying that many expensive sports cars, he would rather put that money back into the business.

Jimmy captivated me. In that first conversation with him, he said, "I want to be the #1 YouTuber of all time. If I can reinvest my money to create bigger spectacles and make a bigger difference, I will do it." He became a client, and later, a business partner. As I've had deeper conversations with him, I've come to realize that MrBeast really does want to make a difference in this world. This is the kind of YouTuber I want to do business with and the kind of person I want to associate with on a personal level.

Planting a Forest of Influence

Fast-forward to the next year, when I was asleep in a hotel room on a work trip. My phone started buzzing off the hook in the middle of the night. I got more than 20 text messages, so I thought there must be some emergency! As it turns out, it was Jimmy, and he was so excited about a big idea he'd just had. He apologized for waking me up, but he just had to share his thoughts. YouTube gives milestone play buttons when channels reach a certain subscriber count: a silver play button for 100,000 subscribers, a gold play button for 1 million subs, a diamond play button for 10 million, and a custom play button for 50 million. Jimmy had done trademark giveaways with each subscriber milestone. He had given his three millionth subscriber three million pennies ($30,000), his four millionth subscriber four million cookies, his five millionth subscriber five million popcorn kernels, and so on up. He was coming up on 20 million subscribers, and he wanted to do something big.

On Reddit, Jimmy saw a "joke" post that said he should plant 20 million trees for his 20 million subscribers milestone. It was on Twitter, too. Jimmy wanted to do it. We talked about making this "joke" a reality, and Jimmy decided to go for it. So he retweeted the post, and it got a lot of positive responses. One of the responses came from another big YouTube creator, Mark Rober. Mark's science-type channel was borne from his engineering and inventing background. He had worked at NASA and in research and design at Apple. Mark wanted to help with the science of the project, and MrBeast agreed.

They created an organization called Team Trees and were excited about making this huge project a reality. In brainstorming and research, Team Trees learned that it actually would be bad for the environment to plant 20 million trees all together the way they had originally planned, so they modified the plan. Team Trees partnered with the Arbor Day Foundation, which is the largest nonprofit organization dedicated to tree planting. The plan was to spread the word and enlist the help of other influencers and organizations to contribute to the Arbor Day Foundation. Then Arbor Day could plant the trees where they were needed around the world rather than all together.

When it was time to make the video about the project, Jimmy wanted to show himself and the team planting the trees, which was to be expected, but I saw the potential to make it bigger than that. This project needed to be pushed by more than just Jimmy with one MrBeast video; he needed to enlist the help of influencers to spread the word and push the movement. We needed to get the word out. So Jimmy reached out to hundreds of YouTube creators to spread the word. Those hundreds answered the call, and they even inspired thousands more to join. More than eight thousand videos were made about the project.

It was widely talked about on Twitter and Reddit. Jimmy even reached out and asked people with big influence (and big money) to contribute to Team Trees. See Figure 8.1. He got the attention of Elon Musk, Tesla and SpaceX CEO, who then donated a million dollars with the message "For Treebeard" (a tree-giant character from J.R.R. Tolkien's classic fantasy novel *The Lord of the Rings*). Elon also temporarily changed his Twitter profile name to "Treelon."

Only MrBeast could give a "wuv u" to Elon Musk and get a million-dollar reply in return! A tongue-in-check response came the next day from Shopify CEO Tobias Lütke, who raised Treelon's donation by one dollar, $1,000,001, and added the message "For the Lorax" (a Dr. Seuss character who "speaks for the trees"). He then temporarily changed his profile name to "Tobi Lorax."

The movement spread like wildfire (pardon the irony), garnering more than a half million individual donations in 55 days. Team Trees raised nearly $22 million. MrBeast tweeted, "We did

Figure 8.1 MrBeast's tweet to Elon Musk

it!! . . . #TeamTrees was more than planting 20 million trees, it was a movement that shows we care and we want to make change."

Raising Money and Awareness

Not long after we reached our Team Trees goal, the Covid-19 global pandemic hit the world. We had just seen the power of big influence, and we wanted to make a difference yet again. So I helped MrBeast produce a live stream campaign with 32 of the biggest YouTube creators of all time to raise money for Covid-19 relief. The live stream was a virtual rock-paper-scissors tournament, and it raised more than $5.8 million. Google matched and surpassed it, donating $12 million. It was the biggest YouTube-sponsored live stream ever. These influencers have the power to really make a difference and help the world in unique and entertaining ways.

It sounds like I'm casually throwing around big numbers here without a second thought. Let's stop and remember that these numbers and these causes are a big deal. Having millions of subscribers is a big deal. Making millions of dollars is a big deal. But using your influence over millions to raise funds and awareness of something that matters and will help the entire world in huge ways . . . that is priceless.

YouTube provides a unique opportunity for anyone to build this kind of following and influence. Sure, many don't choose the philanthropic route, and their YouTube fortune gets squandered away on Lamborghinis and mansions and expensive jewelry for their pets. But some do. Some give money to people on the street or buy an entire grocery store worth of food to donate to a food bank or homeless shelter.

MrBeast has done this more than once or twice. In 2019, he made a video that showed the process of buying all the food in a store and giving it to people who really needed it. In March 2020, when people started hoarding food with the Covid-19 outbreak, MrBeast donated a million pounds of protein. He also partnered with Smithfood, which promised to match every dollar donated with one serving of protein in a campaign called the Good Food Challenge. He reminded millions of people that when you hoard food, you stop donating it to people who also really need it. Just two months later, MrBeast made another video to help people financially affected by Covid-19, giving them money on a fake news station. Some recipients had lost their jobs, while others had large medical bills after contracting the virus. All were humbled and grateful, and some promised to pay it forward.

MrBeast is well known for this big giveaway format. He's given away huge amounts of money for silly things, like "Last to Stop Riding Bike Wins $1,000,000," or, "Anything You Can Carry, I'll Pay For . . ." I took my son Kelton with me on a business trip to work with

MrBeast, and on the flight, I gave him a long "dad talk" about how success comes from hard work and how nothing in life is handed to you so you have to earn it. The next day, Kelton was just observing the video shoot when MrBeast walked over to him and asked, "Kelton, what are you doing?" Kelton said, "I'm just watching you make a cool video." Then MrBeast did something in true MrBeast fashion. He said, "Well, I opened a free bank and I'm giving out free money. Why don't you go stand in line." Kelton was pumped. He got to be in the video, and he got $5,000 for it. Talk about the ultimate backfire. MrBeast literally threw $5,000 in my son's lap after my big speech about hard work with no handouts. MrBeast doesn't do all of this to flaunt the lavish YouTuber lifestyle; he's always working toward his goal of reinvesting his money to "create bigger spectacles and make a bigger difference." It's all a means to an end that matters.

Mark Horvath's YouTube channel, *Invisible People*, exists to change the narrative around homelessness. Mark travels the country to interview homeless people to bring awareness to the problem and hopefully trigger change. Mark said, "We don't need random acts of kindness or a month of impact. We need deliberate, intentional acts of compassion as a lifestyle." *Invisible People* is making a real difference.

Famous people can use their influence to initiate big change. Bill and Melinda Gates started their foundation to facilitate meaningful work in global health and development. The Gates enlist the help of other influential people, including many YouTubers, to spread the word and take action with global issues. Dan Markham from *What's Inside* made a video to highlight the need for clean water in the Philippines and talked about the work the Gates Foundation does. There are so many ways to make a difference in this world, but when you have a lot of people watching you, you can be so much more influential. Take that influence seriously and consider ways you can do something meaningful with your content and your power.

In preparation for the Team Trees campaign, MrBeast told fellow YouTubers in a prelaunch video, "We want to show that YouTube isn't just a drama fest. We actually have real influence and can make real change." This is the type of YouTuber we want our kids to watch and emulate. I will wholeheartedly support my son Bridger's YouTube aspirations if it means he can impact the world like this. If you know a kid who wants to be a YouTuber, you can show them examples of creators who do it in a meaningful way.

9 How Businesses Extend Their Reach and Drive Revenue

Everyone you want to reach can be found on YouTube, whether you're a mom-and-pop shop, brick-and-mortar business, or Fortune 100 company. If you think you can skip or skim this chapter because you're "not a business," think again. Every channel that brings in any money from YouTube or wants to bring in any money from YouTube has to think of itself like a business.

One of my favorite examples of extending reach and driving revenue on YouTube comes from a quilting company in Small Town, USA. Jenny and Ron Doan own the Missouri Star Quilting Company, whose YouTube channel has almost 700,000 subscribers in mid-2020. But in the early 1990s, before quilting was on their radar, the Doans were living in California, and they were down on their luck. One of their seven children had medical issues that required very expensive treatments. The medical bills crushed them. They were nearly bankrupt when they decided it was time to move somewhere more affordable where they could pinch pennies and get out of debt. They literally took out a map of the United States, closed their eyes, and pointed. Missouri was the winner.

So in 1995, the Doans moved to the Midwest to a little town called Hamilton, Missouri. Ron got a job at a newspaper working as a mechanic, but it was a long commute with extended hours and many nights away from home. Jenny found odd jobs to help make ends meet. Unfortunately, they lost retirement savings with the stock market crash in 2008. Ron's job security felt unsure, and the Doans were down on their luck, yet again. Often in a small town, businesses struggle to stay afloat, and this was true for a lot of the shops in Hamilton at the time. Because they were such a small community, it felt like a personal blow every time one of them had to close their doors. The Doans were a big struggling family in a small struggling town.

Demand = Opportunity

Jenny was good at sewing and had been a costume seamstress in California. There was approximately zero demand for a costume designer in Hamilton, so someone suggested that she take a quilting class. Jenny's response? Quilting is for old people. But she did take the class, and what she discovered was that quilting took an incredible amount of creativity. She was hooked. Jenny was a piecer, which means that she put the pieces together for the top layer of the quilt. To finish the quilt, you needed a longarm machine that would put the quilt top, fluffy insides, and quilt backing together. Quilters and their machines were scarce because the machines were very expensive. Quilts ready to be finished had to wait in a long line, and Jenny's quilts were no exception.

One such quilt was finally completed and ready to pick up, and Jenny's son Alan asked which quilt it was. Jenny couldn't remember what the quilt looked like because it had been waiting in line so long. Alan and his sister Sarah were baffled by this. Here was an opportunity for the taking: Mom should get her own longarm sewing machine and become a quilter! The Doans went for it. They invested in the expensive machine, but there wasn't room for it in their house. So they also bought a building. Real estate was cheap in their poor town, and the Doans paid less for the building than they had paid for the quilting machine.

Jenny practiced until she got confident in her skill, and they opened shop. In 2008, the Missouri Star Quilt Company had "launched," but they didn't know who to launch to. In those early days, business was meager in their town of 1,500 people. The UPS guy who delivered stuff to their building felt bad for them because he was sure their business wouldn't make it. You had to be crazy to open a business like this in such a place! One transaction a day felt like a success to the Doans. They made a Facebook page and got two likes. Alan had an idea to do a "Quilter's Daily Deal" to sell random extras kicking around the warehouse, but quilters just weren't online yet. It was a hard start, because they didn't know where to find their audience. The Doans had found an opportunity, but they had invested in an expensive machine in a tiny town with no foot traffic.

Alan enlisted the marketing expertise of his friend David Mifsud to help him run things online, while Jenny and her daughters Sarah and Natalie ran the shop. One thing the Doans did have going for them was their email list. It was important to them that their newsletter bring value to the people reading it, so they gave a lot of helpful quilting information for free. People looked forward to the email and wanted to share it with their other quilting friends, who also would sign up. It was at this point that they made a decision that would eventually transform their lives. Alan suggested to his mom that she should film tutorials and put them on YouTube. Jenny said, "Sure, but what's a tutorial?" He explained it to her, and she said, "No one is ever going to go on YouTube to look for quilting stuff."

Jenny was so nervous and awkward in her first video shoot. Not only that, but she actually had an accident and broke her leg that day! She easily could have quit after such a rough start, but she didn't. She

pushed through the awkwardness of being on camera and continued
to film videos. Traditionally, quilting was an "elite" thing in the world
of sewing, but Jenny made it easy and accessible in her tutorials. She
showed her mistakes and how to fix them. And she genuinely loved
what she was doing. The Doans put links to these YouTube tutorials
in their newsletters, and the readers loved them.

Quilters would get together and have parties to watch Jenny's
YouTube channel. Viewers started asking about the products and
fabrics Jenny was using in her videos, so she told them where to buy
things, and the products would fly off the quilting shelves.

Their website's "Quilter's Daily Deal" offered a 40–100% dis-
count on the item of the day and became a big hit as their reach
expanded, but what really took the company to the next level in
revenue was precut fabric. Finding coordinating fabrics by the bolt
and knowing how much to purchase, what shapes and sizes to cut,
and doing the actual cutting is a huge job. It's the reason why many
people avoid quilting altogether. Jenny's goal was to simplify quilting,
so they made their own line of precut fabrics, already coordinated,
cut, and ready to assemble. They did 14,400 online orders in the first
six months. Business continued to climb, and so did their YouTube
channel. They now do more than 6,000 orders every day, and they
employ half the town of Hamilton.

Missouri Star Quilt Company is the biggest sewing channel on
YouTube, turning Jenny into the quilting world's biggest "sewlebrity,"
as they call her. YouTube helped Jenny reach a worldwide quilting
audience she wouldn't have been able to find in her rural town in
the middle of the country. Her fans even send mail from all over the
world. One letter in particular struck a strong chord with Jenny. In
the letter, a woman from Iran wrote, "You have filled my war-torn life
with color," and Jenny sobbed. She had thought she was just quilting; she didn't realize it could be impacting lives in such an important
way. What she offered might be the only place of peace or happiness
in someone's life.

Think Bigger

In addition to being the biggest quilting YouTube channel, Missouri Star Quilt Company is also the world's largest provider of precut quilting fabric. They even have their own quilting patterns and their own magazine called BLOCK. The company's online reach breathed new life into Hamilton's struggling community. Busloads of quilters visit the town, which has become a tourist attraction. The Doans own and operate 14 quilt shops with different themes, three restaurants, and a hotel in Hamilton, which is now known as "Quilting Disneyland" or "Quilt Town, USA."

Jenny and her family were simply trying to pay the bills when they started quilting, but YouTube gave them a platform to educate and sell to a whole world of quilters. With an estimated $40 million in annual revenue, the Doans are still amazed at how far YouTube took their humble business beyond anything they could have possibly imagined.

Do you understand what I'm trying to tell you? That wasn't just a cutesy story about a nice family who can pay their bills thanks to YouTube. Their annual revenue is in the *tens of millions with a quilt shop in the middle of nowhere because of YouTube*. I do not care what kind of channel you have or what your excuses are, there is a potential to grow and make money on YouTube that anyone can capitalize on! Exactly zero YouTube channels started with millions of subscribers, brand deals, and product launches. Don't sell yourself short by thinking those things aren't for you because you're a small channel. Think bigger. The opportunity to grow in reach and revenue is there for the taking, and if you're not treating it like a business, start doing it now.

Break into Big Brands

In Chapter 2, I talked about how ad revenue sharing changed everything for the YouTube ecosystem. I introduced Orabrush as an original case study of leveraging the power of YouTube to generate huge success. Go back and read about how Jeffrey Harmon, a poor college student at the time, took a product from near death to international distribution and millions in evergreen sales. Orabrush's inventor, "Dr. Bob" Wagstaff, had tried many ways to sell his tongue-cleaning brush, but nothing had worked. When a company is spending two dollars to make one, they get to a point where they have two options: quit, or try something totally new. Wagstaff chose option two and gave this marketing student the reins. YouTube was totally new, and Jeffrey wanted to try using it to help Dr. Bob save his invention. Advertisers weren't on YouTube yet; it was a rudimentary platform to buy ads to send traffic to. This is why it was such a huge opportunity: there was nobody to compete with yet. Jeffrey enlisted the help of fellow amateur creators he knew to help him make a video for Orabrush for a few hundred dollars, and he put it online.

Jeffrey paid a penny per ad view—a crazy cheap amount by today's standards—and Orabrush started making big money. The ad platform had increased sales percentages by the *thousands*. In fact, Jeffrey kept buying more ads and making so much money in those early days of ad sharing that YouTube had to scramble to figure out how to cap advertisers' revenue percentages. Because of one guy! He had been buying up all the inventory in ad views, and it was like printing his own money. YouTube caught on, and they wanted to control the money flow. (I like to call this moment in YouTube history the Jeffrey Harmon Effect. Oh, the power!)

In the ad's first month, Orabrush made $30,000. The second month saw $70,000 in revenue. In 2009, Jeffrey helped Orabrush create a web series called "Diary of a Dirty Tongue" complete with a man, Dave Ackerman, dressed in costume as a giant talking tongue. The idea behind the web series was to create brand followers, which was unheard of at the time. Jeffrey had Dave and the tongue costume join him at VidCon, attending as the convention's first sponsor. There were 600 attendees that year. (As a reference, VidCon US 2019 brought in more than 75,000 attendees. And it was only one of several annual VidCon events around the world.) Jeffrey's takeaway from the conference was the realization that everyone at the event also wanted to build a following.

They continued to create episodes for the web series, and they had more than 200,000 followers. Orabrush was the first brand to create a web series; nobody had done that before. It was definitely ahead of its time—Jeffrey even admitted it was maybe too far ahead of its time as far as achieving reach and revenue potential goes. The world didn't know how to do that yet. Back then, even respectable cable networks only had 100,000 subscribers. The biggest YouTube creators at the time had the same amount of followers, which meant that these individuals were basically their own cable networks. The difference was that they didn't have to answer to any executives. This was an eye-opening realization. Jeffrey was beginning to understand the potential power of leveraging YouTube for massive growth.

Orabrush had created true fans and had generated millions of video views because people shared the ad organically online. Their affinity with their customers made it possible to get their products into Walmart, Costco, and international markets.

The second year Jeffrey went to VidCon, a guy from Google approached him and gave him an actual hug because of Orabrush. He told Jeffrey that they had been unable to break into big brands, but because of what Orabrush did with Walmart, he had been able to land a deal with Coca-Cola . . . and probably saved his job. Orabrush had changed the trajectory of how business was done. "When you figured out how to make money," the guy said to Jeffrey, "it opened the door for everyone."

Optimize with Split Testing

Next up for Jeffrey and his brothers Neal and Daniel was a campaign with a company called PooPourri. The Harmon brothers had begun working together as an ad agency without making it official. The ad campaign for PooPourri was a quick one, going from inception to launch in only three weeks. In just 21 days, they did the script writing, casting, shooting, editing, and ad release. That's impressive. The ad truly went viral. When people talk about a viral video, they are talking about ads like the one for PooPourri. Because of what they had learned doing Orabrush, the Harmon brothers had learned how to do split testing with different intros, outros, and lengths. In testing, William Goodman, a *Huffington Post* reporter, got his hands on the video, and he embedded it into the article he wrote about the ad called "Poopourri Spray Promises to Take the Stink Out of Public Pooping." The article made the video go viral long before testing was complete, so Jeffrey and the client decided to go ahead with the campaign launch right away.

As successful as the ad was, Jeffrey said it could have gone even further and generated even more revenue had they been able to finish the split test. The video that had gone viral was titled, "How to make it so your poop doesn't stink," and testing showed that the title, "Girls don't poop," was performing better. Here's an expert tip from Jeffrey: use the words your followers use to describe your brand. "Girls don't poop" had been said several times in the comments of the video, so they took the phrase and turned it into the best performing version of the ad. "You can find your true brand message in the comments," Jeffrey said, "If you aren't using those words, you're probably not on target." Take it from Jeffrey—he's run a successful ad or two. The Harmon brothers made a very large amount of money on the PooPourri campaign. (Fun fact: PooPourri asked where to wire the money, and the Harmon brothers didn't even have a business account at a bank yet. They hurried to file as a business on their state's website so they could run down to the bank and set up an account.)

Take Smart Risks

Another client the Harmon brothers helped generate a ton of views and revenue for is Bill and Judy Edwards and their son Bobby, owners of Squatty Potty. Squatty Potty is a footstool made to use while going "number two." It is ergonomically shaped to fit perfectly under a toilet for easy storage. Squatty Potty had been on the TV show *Shark Tank*, a reality show that gives entrepreneurs the opportunity to pair with an investor and get their product off the ground. A *Shark Tank* investor signed on with Squatty Potty as a 10% owner in the company.

The Edwardses had gotten their toilet stool into stores, but people didn't know how it worked. They thought it was a gag. Soon, the revenue Squatty Potty made with the investor plateaued, so Bobby Edwards needed to penetrate the market more than was currently being done. He needed to get the product in front of more people. He asked the Harmon brothers to help, but he hesitated when he heard their pitch for an ad with a pooping unicorn. Bobby's investor said no. The investor said it should be a free campaign, which meant the Harmon brothers would be doing it pro bono. It was Jeffrey's turn to say no. He knew how valuable their skills were, and he had the résumé to show for it. They agreed to disagree, and Jeffrey went on his merry way.

Fast-forward three months. Bobby went to an event where there happened to be unicorns in attendance. It must have been a positive experience, because Bobby came back and told Jeffrey and me (I was executive producer on the project) that he wanted to do the ad. He gave the executive go-ahead to the Harmon brothers without involving the investor who had said no. To Bobby's credit, this was a huge risk for him—the campaign was very expensive . . . and he had a 10% investor to answer to eventually.

I met with Jeffrey, Daniel Harmon, and Dave Vance, a brilliant writer, at a retreat to hash out the creative ad the campaign.

We all brought our own version of a script, but Dave Vance's was the clear winner. It featured a unicorn and a prince, and it was magical. We paired Dave's clever script with the sales components from Jeffrey's script and my script. Then we let Daniel work his artistry on the creative side. Together, we had created something special. The retreat gave us the uninterrupted space to plan a killer campaign.

We went home and started production right away. There were kinks to work out (no pun intended . . . go watch the ad), but we kept the campaign plan intact and moving forward. During filming ads in the past, we had run into issues with clients being on set because it stopped the flow of creative work being done. A word of advice when you're in the same situation: let the creatives create; don't let the brand dictate. So we planned to film the ad while Bobby was out of town at a convention with Squatty Potty to avoid this issue. Bobby and his parents worried about being gone when we filmed, but we told them, "You hired us, so trust us."

A few weeks before launch, Bobby had been hounding me to send him the ad, and I didn't know what the big deal was. Why was he making it sound so urgent? Finally, he told me that he still hadn't talked to his partner about his decision to move forward with the campaign. He had invested a lot of money into it at this point, so if the investor rejected it again, he would be out a lot of money. We sent him the ad so he could show it to the investor . . . and they approved this time. The aggressive potty humor in the ad had the possibility of being criticized, but it was so good that it was worth the risk. We decided to launch on Facebook, which was a little nerve-wracking for two YouTube guys. Facebook ads were still pretty new at the time, and we thought it would perform best there before we launched it on YouTube.

We weren't wrong. The ad went viral. The campaign on Facebook and YouTube got more than 20 million views in its first day, and we hadn't put a single penny into ad spend yet. We made a lot of money. Again, the big platform had to scramble to figure out how to regulate advertiser revenue because of Jeffrey Harmon and his team. Facebook now has regulations to cap ad revenue, but there is still so much to be made. This ad campaign was only half a million dollars, and we got our ROI in just a few days. Squatty Potty went from having a couple million in revenue to $28 million from one campaign.

An adorable pooping unicorn and an eloquent prince delighted millions of people and drove tens of millions in sales. Comments, tagging, and sharing skyrocketed sales, and the product flew off the shelves. I had warned Bobby to increase inventory production to satisfy the demand we knew would be coming. He did increase some, but conservatively. He had already invested so much into this campaign. Nobody could have predicted how far the ad reached and how many people wanted to buy—it was crazy. Bobby ramped up to a 24/7 production right away, but even with manufacturing based in the United States, he ran out of Squatty Potty stools before the Christmas rush.

Every brand launching an ad campaign should be concerned with two things primarily: the brand's message and split testing. Stay true to messaging at all costs. And split test to see what works best for optimal results and revenue. When split testing for Squatty Potty, we found that the ad with the $29.95 cost actually performed better than the $24.95 ad, so we bumped up the price to $29.95. It converted better and made more money for everyone. Win-win. You also need to think about distribution ahead of time. With its humor, we knew this ad would be shared organically and that we could amplify it with a paid strategy. We knew that even if an ad platform cost us more, we could get the revenue we wanted.

Think Evergreen

If you are a business, you need to stop thinking quarterly. Make content that is evergreen, meaning it will continue to produce results year after year. No platform likes old ads, so shoot at the three- to six-week zone for initial results, but any piece that has the fundamentals of good ad content down will continue to work. There always will be people who haven't watched it who will see it like it's brand new. The original Squatty Potty ad aired in 2015, and it is still running today. And every time they rerun the Squatty Potty episode of *Shark Tank*, we see a spike in sales.

Be Your Own Influencer

YouTube has been built around creators who become their own brand influencer. This is not the Hollywood way. Hollywood has two problems: the studio sitting between the audience and the influencer, and the sheer amount of content. There is just so much content to consume. Netflix, Hulu, Amazon Prime, HBO Max, Vudu, Disney Plus, Apple TV, Peacock, and so many more are all vying for consumers' attention. How can your brand cut through all that noise? Build a following that has an affinity with you as an influencer. With a hundred hours of content being uploaded to YouTube every second, that's the only way to get yourself seen. You must have a direct connection with your audience.

If Hollywood wants to survive, they'll have to adopt this model, too. Actually, it's already happening. Dwayne "The Rock" Johnson, for example, commands a bigger paycheck, not because of his acting ability, but because of his massive social media following. He has connected himself directly with his fans, and they love him for it. Where they used to be three separate things, creators, influencers, and brands are all the same thing now. Jeffrey said it best: "Weak brands piggyback off of influencers. Good brands *are* the influencers." This is how you extend your reach. Connect with your audience directly, and watch the money follow.

Embrace the Digital World

Companies that hold fast to traditional marketing only are stuck in the 1990s. For so many of these companies, egos and awards continue to get in the way of actual money making and brand reach. They'll run "studies" or fluffy ad campaigns to pander to executives, shareholders, and investors. And in return, they'll get a pat on the back and a little more job security. These companies don't look at real numbers to measure campaign revenue and brand lift, and they don't think about direct-to-consumer e-commerce that could be raking in millions more.

Companies that embrace the evolution of marketing are way ahead of the curve. They have the right framework in place to drive conversions. The right framework means they keep traditional elements like problem/solution, entertainment, and brand awareness, but also they integrate newer strategies that work in today's digital age. Strategies like direct response, attribution, and influencer activation.

There is so much available to help you track marketing these days. Analytics, pixels, and brand lift are huge metrics to measure. You can do an at-home brand lift study by going to Google and looking at your search increase and direct traffic increase. You can look at Google Trends as well. If you want to spend the money, Google can do a brand lift study with you.

Sometimes these numbers get buried because companies and creators just don't know how to look at them right. A company might look at their in-platform tracking pixels and think their ROI is only two-to-one, so they bag the campaign, labeling it a failure. This is where omnichannel attribution must be considered. Omnichannel means a customer's experience with a brand should be integrated and seamless among online, offline, social, and mobile interactions. And that businesses need to be able to track all of those streams of revenue.

In 2013, ad agency Crispin Porter + Bogusky (CP+B) had hired 25-year-old Travis Chambers as a social media supervisor. That same year, the big CP+B execs had a meeting with Turkish Airlines, whose next campaign goal was to have the most viral ad of all time. At the time, CP+B was at the height of their success as "Agency of the Decade." CP+B's Chief Digital Officer Ivan Perez-Armendariz was in the room, and he privately texted Travis to "Get in this meeting." There were a lot of executives and creative directors/teams in attendance, but Perez-Armendariz knew that they needed Travis's specific skill set on this project. Travis told them how they could run a successful campaign, and he landed the job of distribution and content strategy for the Turkish Airlines campaign.

Perez-Armendariz gave Travis the green light to do what he needed to do for the campaign to be what it needed to be. Involving too many execs and too much red tape would bog down the creative freedom needed for the project to be as successful as it could be. So he let Travis run with it. Travis hired Ricky Ray Butler's brand integration company Plaid Social Labs to connect influencers with the campaign. They got professional basketball player Kobe Bryant and professional soccer player Lionel Messi to compete for the attention of an adoring kid fan in the ad titled, "Kobe vs. Messi: The Selfie Shootout." CP+B had been tasked to get more than a 100 million views on the campaign in order to beat the previous year's commercial. Travis, along with Plaid Social Labs's help, had 800-plus influencers tell their viewers to check out the video by Turkish Airlines.

Their communities embraced it and shared it. It got more than 140 million views. The total campaign cost about $3.5 million, and—here comes the crazy part—they didn't even track it.

Nobody at CP+B looked at the numbers, except the president of the Europe office, who sent an email detailing his shock. He compared the YouTube campaign's $3.5 million ad spend to their traditional TV buy, which cost $25 million and got half the impressions the YouTube campaign got. He wondered why they tracked data and brand lift on the TV buy but not on the more successful YouTube buy. And he suggested that the YouTube strategy should be the agency's game plan moving forward. Nobody listened. They were busy getting Cannes Lions and Grand Prix awards and patting each other on the back. The ad was dubbed the "Most Viral Ad of the Decade," and nobody even knew exactly how much money it had made. Travis's guess is that it was easily more than $50 million across all streams of revenue.

Apply Omnichannel Attribution

Whether you are a small business or a corporate giant, you must be doing omnichannel attribution. Omnichannel attribution tracks and measures a customer's overall interactions with a brand across online, offline, social, and mobile experiences. It tracks streams of marketing revenue, so brands know what flops and what works so they can do it better (and make more money) on subsequent ad campaigns.

For example, Travis went on to create his own ad agency, Chamber.Media (and eventually land on the 2018 Forbes 30 Under 30 list), and he ran a campaign with Sephora and PMD Beauty. Sephora mandated that PMD increase retail sales by 15% in order to remain in their stores. Without omnichannel attribution, Sephora would have thought that the campaign had only gotten a one-and-a-half-to-one return, but Chamber.Media went and pulled in all retail plus other revenue channels in addition to YouTube, Facebook, and Google ads, and they found that it was actually a blended four-to-one return. They raised retail sales by 30% and saved PMD from losing Sephora.

Another Chamber.Media client, MRCOOL, thought their ROI was only two-to-one from the in-platform tracking pixels, which is pretty bad for a $1,500 average order value product. But when they took in homedepot.com, lowes.com, their parent company who also sells on their website, and Amazon and Google shopping, it showed a 20-to-1 return. The few million dollar ad spend had driven $50 million in revenue. "It's one of the most successful campaigns we've ever had," Travis said. "And someone could have turned that off if they didn't understand the omnichannel attribution model.

So many companies are missing the big picture here, turning off campaigns because of poor tracking, or doing a short-term campaign instead of thinking evergreen. You have to get tactical. You have to look at all the data and plan long term. Television campaigns are very difficult to track. Digital campaigns can be tracked, measured, and are evergreen. If you aren't using omnichannel marketing and tracking, start doing it now.

Use Paid Acquisition

Travis and I have worked on several projects together, and we have seen the power of YouTube to help businesses expand in both reach and revenue in ways they'd never dreamed. Organic growth may be slowing down from the gold rush days as the Internet saturates over time, but there are always ways to be tactical and creative.

People like the Harmon brothers and Travis Chambers got in when the getting was good in free reach. It used to be that you could get a lot of free reach with audience growth, a social following, press, affiliate marketing, SEO rankings, and influencer deals. These efforts are difficult to grow sustainably for a long period of time now, because Google, Facebook, Pinterest, TikTok, Amazon, and others own the real estate in reach. That initial "gold rush" is heavily mined, so the way to get seen now is best achieved with paid acquisition strategy, or ad buying. And as long as your content is creative enough, you should be able to sustain.

Ad buying is probably going to get extremely easy in the next few years, because Internet algorithms and artificial intelligence will be able to do the job of ad buyers. Where you'll have a leg up on the competition, then, will be with awesome, creative content. Robots will never be able to think outside the box like a human can. So, if you really want that reach, make sure you have a great creative plan in place, and use a paid acquisition strategy that will last forever. A brand that's doing this well is Purple mattresses. They use paid acquisition, but they have incredibly creative content to back it up.

Other brands that explain a problem and solution clearly, even without words sometimes, perform really well. They hook visually, which works wonders in a world that's endlessly scrolling. It takes a lot to catch consumers' attention these days, especially in a sea of endless content. So if you can grab their attention in the amount of time it takes to scroll, you must get to the problem-solution point as quickly as possible. Instagram ads are really good at this. Pay attention next time you're on there.

Learn Strategies and Use Them

It doesn't matter what kind of channel or business you have, anyone you want to reach can be found on YouTube. Learn how to leverage YouTube's power to connect you with your audience. Use strategies from people who have done it successfully. It can feel too difficult to cut through the noise of so much online content these days, but it can be done with the right creative and paid strategies. Learn them, and use them. Don't make it guesswork. Utilize tracking systems, ideally with omnichannel attribution. And implement the data-driven approach to YouTube success coming up next in Part III.

PART III The YouTube Formula

10 The Data-Driven, Human-Centered Formula

It can't be that hard to start a YouTube channel that gets lots of views and subscribers. That teenager makes videos of himself goofing off with his friends, and he is a millionaire. I can do that. I'll just hit record, click upload, and watch the money pour in!

Unfortunately, I've seen way too many YouTube creators who started their channels thinking it would be easy. It doesn't take very many videos with little to no views before a creator gets frustrated. YouTube is harder than they thought it would be. We've all been there. You took a lot of time to plan, film, and edit, and you think you made a pretty darn good video . . . so why doesn't anybody want to watch it? Turns out the YouTube swimming pool is actually an ocean, and you're just a tiny fish in the sea. I'm always impressed when people keep grinding on YouTube without getting the results they want. It's super frustrating, but they refuse to give up.

I love YouTube, and I love helping YouTube creators and businesses. I'm lucky to work with all kinds of channels all over the world. No matter how big or small the channel is or what type of content they create, every YouTuber wants me to help them with the same thing: they want to grow. Some want to grow in views, some want to grow in subscribers, and some want to grow in money. Actually, different types of growth usually go hand in hand. When you get more views, you get more subscribers, and income growth generally follows.

Part I of this book was to help you understand YouTube, the history, the ecosystem, and the platform. Part II was to help you discover the possibilities and opportunities for creators and businesses on YouTube. I hope you were educated and inspired. I hope it made you reconsider your Why. Part III is where I will help you learn how to get smart and create content tactically. I want the wonderful opportunities for success to be yours, too. I want your hard work and passion to pay off. YouTube has given creators some amazing tools and data to help us figure out how to reach more people with our content. You probably won't see YouTube's success secrets in a rain of digital code unveiled like Neo did in *The Matrix*, but you'll learn the step-by-step Formula to help you achieve the YouTube success you've been dreaming of.

If you want your videos to get discovered in a sea of millions, keep this thought at the forefront: make data-driven decisions but always optimize for humans. You have to become a YouTube analytics pro, whether you like it or not. But this doesn't mean you turn into a robot or create content for a robot. Your YouTube success hinges on you maintaining the human element. You have to create for humans—the robot was built to find and follow the humans that would be the most likely to love your videos.

YouTube Sixth Sense

Dan Markham is a very successful YouTube creator. One day I got a call from him asking if he could use my studio to film a video because he had recently moved to the area where I live, and his studio wasn't set up yet. Dan and his son Lincoln came over and filmed a video for a couple of hours. On their channel *What's Inside*, Dan and Lincoln cut stuff open to see what's inside (a brilliant idea for a channel . . . why didn't I think of that?). That day's video featured the cutting open of a real Rolex watch and a fake one. You can find the video by searching "What's inside REAL vs FAKE Rolex."

They got done filming, and I thought it was a pretty good recording session. But Dan turned to me and said, "Well, this video isn't going to do very well." I was surprised. I asked him why he thought that, and Dan listed several things that would have helped with audience retention in this video. Audience retention is an important metric that tells if the viewer actually enjoys the video. I asked if he thought he could fix it in editing or voice-over, but he said no. There were some elements that were needed to make it happen in real time and editing wouldn't work. And he couldn't refilm the video because he couldn't cut open and destroy another expensive watch. Dan did what he could and uploaded the video to his channel, and guess what? His intuition was spot on. The video underperformed exactly like he said it would.

How did Dan know that his audience wouldn't respond to this video? Does Dan have some freakish sixth sense?

I have found that the most successful creators have great intuition about what will work and what won't. They know their audience so well that they can kind of predict the outcome in their head. They know which video will become a "banger" and which won't.

This kind of YouTube Sixth Sense can be learned, and it comes from analyzing successful content and knowing your audience. To analyze content, you have to put it under a microscope and figure out every moving part. Data analysis will force you to move past biases and assumptions that can cloud judgment and lead to poor decision making. Becoming a data expert is what helps you discover your own YouTube Sixth Sense so you can intuitively know how your content connects with your audience.

The perfect example of a creator who has this YouTube Sixth Sense is Jimmy Donaldson, "MrBeast." I've been working with Jimmy for a few years, and I've never seen a creator like him before. Never. He has been disrupting YouTube and the Internet with his "super power" for successful content creation. Was he just born with it? No! He watches YouTube differently than most. He imagines how people will feel and respond to every aspect of the video and makes decisions from there. He really thinks about what will get them to click and when they might lose interest in the video. But more importantly, he validates his intuitions with data. He has always been obsessed with data. Jimmy and I first met in person because he wanted to have a face-to-face conversation about data without interruption. He literally dropped everything and got on a plane just so we could talk about YouTube data. His data obsession is why he gets tens of millions of views every day and millions of subscribers every month.

You have a unique perspective and personality that can connect with a specific audience. This audience just needs help finding your content in a really big ocean, and the way to do that is by making data-driven decisions. Instead of getting frustrated when nobody watches your video, get tactical. Take control by learning the trade in numbers and graphs. The YouTube AI is always actively looking for a potential audience for your content. YouTube has given us the tools so we don't have to rely solely on our gut feel or intuition.

Know Your Goal, Know Your Why

Every creator and business should be able to answer the question, "What is your finish line?" or, "What is your goal?" You might be surprised to learn that many creators can't answer this question outright. If you don't have a goal in mind, you're probably running in circles on YouTube. I want to help you figure out your finish line so we can make a game plan to get there. This is the first step in the YouTube Formula.

An easy metaphor here is running a marathon. Marathon runners have a literal finish line to cross, but their goals can range from running the whole thing without walking, to beating their best time, to qualifying for the Boston marathon or the Olympic trials. Each runner in these scenarios has to run the exact same course for the exact same distance, but their training plans will look wildly different from one another because their goals are different. Whatever your finish line is, you have to define it in order to make a plan, know the steps it will take, and how to track yourself to get there. Most creators want to get millions of subscribers and dream of the day they can post a picture of themselves holding their own gold play button. Your goal could be as simple as replacing your nine-to-five job with your YouTube income. Maybe you want to influence change in other people's lives. Or maybe you're like MrBeast and want to be the biggest YouTuber of all time.

Everyone is motivated by different things. Personally, I'm not motivated by rewards like more views, subscribers, gold play buttons, or money, but I am obsessively motivated by goals. I feel so much accomplishment and satisfaction by achieving a goal. Maybe your motivation is money, and that is totally fine. We just need to know that right out of the gate so we know what plan to create.

Let me tell you about a YouTube creator from the channel *Invisible People*. Mark Horvath interviews homeless people on the streets. His channel has more than half a million subscribers and more than 120 million views. I asked Mark why he was on YouTube and what his finish line looks like, and he told me he was on a mission to help educate the world about homelessness. He explained the misconceptions about homelessness that make it really hard to address the root of the problem. Mark then told me a heartbreaking story about a kid named E.J. E.J.'s parents died in a car crash when he was six, so he was placed in foster care. Foster care was a nightmare for E.J., who was abused by his guardians, so he ran away. He lived on the streets for 11 years.

Mark said to me, "Derral, people just don't get it. Politicians don't get it. And I'm on a mission to let people know there is more to homelessness than meets the eye. I want to change the narrative." I could see that this meant so much to him, so I asked him why. Tears welled up his eyes, and he told me that he used to be homeless. He was unemployed for an extended period of time, which caused him to lose his house to foreclosure, and he had nowhere to go. He felt like he became invisible when he was homeless. This is why Mark was so personally passionate about the reason behind his YouTube channel. He said, "This isn't just talk. Each year, our groundbreaking educational content reaches more than a billion people across the globe. Our real and unfiltered stories of homelessness shatter stereotypes, demand attention, and deliver a call to action that is being answered by governments, major brands, nonprofit organizations, and everyday citizens just like you." Mark's finish line is perfect for him.

When I work with clients and students, I help them create a plan to follow the YouTube Formula and take the steps to get to their own finish line. The Formula will help you create any plan to fit any goal for any channel, because it is formulated from the data YouTube gives us. YouTube is awesome at giving creators data that will help them make smart decisions based on real feedback; we don't have to guess what to do next. Know your goal, create a plan, execute it, analyze how it was received, and adjust your approach moving forward. In short this is the YouTube Formula, but there are a lot of moving parts and data we need to understand to really take advantage of it.

Become Aligned with YouTube's Goal

As I explained in the book's introduction, my career started in 1999 when I started my own online marketing company. I did website ranking on Internet directories and search engines. As I was learning how to do Internet marketing in the 1990s, I had learned a lot about search engines and SEO. Basically, I was good at keyword stuffing and looking for exploits. The Internet was dumb back then, so this worked. Traditionally, search engines had been done by submitting to a directory, but Google changed the way everything was discovered by spidering the results and then ranking in search results. Page one of Google search results is prime real estate in money land, and obviously, everybody wants it; while page two of search results . . . well, if you've never heard the adage about the best place to hide a dead body . . . then google it (spoiler alert: the answer is page 2 of Google).

I spent years trying to find exploits, using hacks to get to the top of page one for my clients. Every time I thought I was on top, Google kept doing updates to plug the holes and figure out Internet traffic, and they would give every update a cutesy animal name: the Panda update, the Penguin update, and so on. I hated these updates because they made my job and life so difficult. Then I would be back on that cycle, looking for any hack or strategy to get those websites to rank. All I could think about was the algorithm. This was a vicious cycle, and something had to give. What gave? I gave. I finally realized that I had to change the way I did things so that I could operate in line with Google's goals. Once I did that, I succeeded every time.

This is my plea to you not to hate the YouTube algorithm. Take a step back and realize that the AI's job is to follow the viewers and predict what they will watch, and *this is a good thing*. YouTube is not out to get you! Don't try to game the system like I did for too long in those early Internet days. Think of the viewers as people, real humans living and breathing. YouTube is built to help humans find the videos they want to watch, videos they enjoy and are satisfied with.

This is why you have to optimize for humans, not search engines. To be the best creator you can be, you have to be making content for people. So many creators put too much importance on keywords, thinking they are the keys to the YouTube kingdom. Let me tell you the truth: YouTube looks far beyond your keywords anyway. When you upload new content, YouTube has systems in place that determine exactly what is in your video.

The YouTube AI doesn't care about your title and thumbnail, but if a person cares and clicks, then YouTube pays attention. When the viewer connects with the content or the creator, they watch more and they watch longer. This makes YouTube happy, and they push your content to similar viewers. Most people push content, content, content, but they need to be more worried about connection with the viewer.

The first time someone watches your content they are clicking because of your title and/or thumbnail, yes, but if they watch the video and start to connect with you, then that's your golden ticket. When you convert a viewer to a fan, your title and thumbnail lose a lot of importance because that fan will watch your video anyway at this point—they like you and trust you, and they want to watch based on those facts.

In order to optimize for humans, I recommend that you define your video with a title that has fewer than 60 characters. This is hard to do, and it will take more thought than you realize. The human eye is attracted to shorter titles. Humans, as opposed to robots, are attracted to certain pictures and have an amazing capability to browse through them fast. A human brain can process thumbnail images in a jaw-dropping 13 milliseconds. And once the brain lands on an intriguing image, it takes an average of 1.8 seconds to process the attached title.

Additionally, humans respond to storytelling. Your video should have a story arc. It should begin with you delivering on the promise of your title and thumbnail. After that, you need to reengage the viewer and pull them into your content. This is a great time to share the personal touches: your backstory, your beliefs, and other engaging elements. Doing things like sharing your beliefs puts your viewers in a position to decide whether they agree with you. At the end of the video, you again deliver on the promise of the title and thumbnail, proving that they got what they came for, and you tease a follow-up video to keep them watching longer. Finally, you interact with your viewers and respond to comments.

Once you try to optimize for humans instead of search engines then you will do exactly what the algorithm wants, and YouTube will be able to find your audience. The AI looks at patterns for people, so when you create content for people instead of for robots, everybody wins.

Traffic and Momentum: Decoding the Viewer

I first learned a very important principle when I started my company around the turn of the century. My clients were often the mom-and-pop type local businesses, and they were difficult to convince that spending money for online marketing was worth it at the time. I mean, we were still on dial-up Internet. Do you know the dial-up sound? I can still hear it in my head to this day. It took forever to connect to the Internet! Anyway, one of my first clients came to me for help with their marketing because they had tried TV, radio, newspapers, and other resources, and nothing seemed to be helping. I convinced them to let me do some online marketing for them.

It took a couple of months, but customers started coming in their door "magically." These people were coming in because of my online marketing efforts, but my client didn't recognize the correlation and didn't ask the customers where they had heard about them. They told me they no longer needed my services because business was booming. I was devastated to lose their business. They didn't take the time to figure out why the traffic was coming in the door, but I knew. I also knew their traffic would slow down when we quit our online marketing, which it did after a couple of months. They came back to me and we reengaged, and they ended up growing by leaps and bounds.

The important principle this client had missed was knowing how to replicate what had worked. While I knew the Why behind their results, they didn't, and they naively assumed they didn't need to know. You have to know why your traffic came and where they came from. A lot of businesses have made bold assumptions about things they thought would work when the complete opposite was true. You have to know which message works for which traffic sources.

Some of you are brand new and some of you are seasoned YouTube creators, but all of you need to know that traffic source is the key to understanding your viewer and their behavior. Here's why: if you go to the data that YouTube gives you, you'll see a list of where your viewers are accessing your content from, including YouTube Search, Suggested videos, Browse features (Homepage and Subscription), Playlists, Channel pages, YouTube Advertising, Videos, Cards, Annotations, Notifications, Endscreens, and other features. You'll see how many views are coming from where, and you'll notice that the other metrics are completely different depending on which traffic source you are looking at.

If you remember in earlier chapters when we discussed the AI, you'll remember that there are multiple algorithms with an "s." This is where the "s" comes into play. Each traffic source has an independent algorithm whose aim is to increase the likelihood of people who will click and watch videos. Knowing where your traffic is coming from can change the way you do everything.

Kristina Smallhorn is a realtor from Louisiana who was using YouTube to create content from a Search-traffic mindset. She spent so much time on keyword research and creating content for Search traffic. She did see some positive results with clicks and views, and she got leads and closed some deals from her content, but she wanted to improve and to grow her income even more. She had bought a few online courses and read some books on SEO, and they just weren't getting her the results she wanted to see. At this point, Kristina reached out to me for help in a consultation.

I looked at Kristina's channel data and we talked about which videos had the best numbers that were still bringing in views. There were two candidates: "How to Get Rid of Dog Pee Smell In Carpet," with 223,000 views, and "Mobile Homes Pros and Cons: Manufactured Homes," which had 107,000 views. After digging into the data, I could see her older videos in particular had performed well, so I recommended that she make more content to reach the viewers that had responded to those two videos.

She pushed back on my first recommendation. She didn't want to become the "Dog Pee Lady" on YouTube. We both chuckled about that. However, she was willing to give mobile homes another try, so she went and did exactly what I told her to. She didn't have a lot of faith that it would work, so when it did, she was thrilled. Now, her channel growth stats are off the charts and climbing. She was able get six million views and more than 33,000 subscribers all because she was able to make content for a specific viewer that YouTube could find and recommend her content to. The best part was that she was also able to get a flood of new leads for her real estate business and more referral income, and she was able to find other income streams that would replace the need to be a listing agent.

If Kristina hadn't known to look at her real-time traffic and traffic sources to know what content to make and how and for whom, she would still be spending tedious hours on keyword SEO. Know your traffic sources so you can know your viewer.

I understand if you are frustrated. I understand if you have thought about quitting YouTube. But I also understand that if you stick with it, knowing your goal and learning the Formula, you can achieve everything you want, and so much more. Get ready to dive into Part III with this mind-set, and complete the Action Exercise tasks at the end of each chapter.

Action Exercise

Task 1: Write out your finish line (your goal) and put it somewhere you will see it daily. Don't lose sight of why you're creating, especially in the hard times.

Task 2: If you have content on YouTube, go to your real-time analytics and look through your top-performing videos. Write down the ones that are more than 6 months old that are still bringing in the most traffic in the past 48 hours.

Task 3: Plan, create, and upload a video about the same topic as one of those on the list you made.

There is a free companion course that goes with this book. You can get free access to all the videos, extra training, resources, and workbooks at www.ytformulabook.com.

11 Identify Your Audience

Most new YouTube creators use the classic "spaghetti approach" to make content: they throw it against the wall and see what sticks. This might work if you want your pasta just right, but it does not work for successful YouTube content. You can't upload content willy-nilly and cross your fingers that everyone will watch it; you have to get tactical to figure out what works with a precise audience. Your audience consists of the individual viewer, which we sometimes call the avatar, or the representation of the actual person on the other side. I use these terms interchangeably throughout the book. Understanding the avatar is the most important thing you can do. It doesn't matter if you're using video to grow an audience or to sell something, you have to know who your audience is.

Before you worry about the audience, though, you must focus on your content. If you're not making good content in a niche you are passionate about, your channel will fail, plain and simple. Do something you care about, and make good content. Period. The audience will follow, and YouTube will follow the audience. I will show you exactly how to get tactical about it, but remember that you have to start with a commitment to a passion. I have seen plenty of creators who change their content too quickly; they aren't giving the content a chance to sit long enough to gather data and make a strategic plan.

If you are working with something you care about, you should be willing to put in the time to let the data help you.

Don't Make Videos for Yourself

Once you land on a content foundation you're passionate about, you are ready to pinpoint your specific audience. Here comes the big question I have heard thousands of times: How do I find my audience? It is not a dumb question; in fact, it's the best question to get an answer to if you want to be successful on YouTube. Even creators who are doing well sometimes don't have this question answered, and it's holding back their channel from really taking off. Shaun McBride, "Shonduras," was already a successful social media star when he started his YouTube channel, so he had an inherent audience, but his channel didn't really find its stride until after he had made more than 800 videos. "Once I focused on a certain demographic and was true to who I was and what I wanted to film—life with my family outside of work—that's when I found a loyal audience and saw massive growth," Shaun said. I don't want it to take 800 videos for you to identify your ideal target viewer. That's why I wrote this book: to help you plan, execute, analyze, and adjust your content based on viewing patterns so you can find your audience faster.

Create content around your passion, but don't do it for that reason alone. Do it for the people who will actually watch your videos. Combining your passion with your ideal audience creates a crossover that works like a charm.

The good old Venn diagram shown in Figure 11.1 nicely lays out what I'm trying to say here.

Stay true to what you want your channel to be about, but don't make the video for you, make it for the audience who cares. If you don't find this crossover between your passion and your audience, your channel will never grow.

Figure 11.1 Valuable content

Chad Wild Clay and Vy Qwaint are two creators who are pros at figuring out who their audience is and what they'll watch. They used to do parody videos, and one of them went viral in response to the Pen Pineapple Apple Pen phenomenon. As a result, they learned that viral videos don't necessarily do you any favors. They got a lot of subscribers from the video's success, but they soon realized that a subscriber doesn't usually equal a dedicated viewer. There is a huge difference between building an audience off of a viral video and building an audience off of your content and your personality. You want a more stable viewership; you want viewers who are loyal to you. The viral video doesn't attract the same type of viewer as one who would loyally follow regularly scheduled content and personalities.

Something Chad and Vy did right was read their viewers' comments and create new content from suggestions viewers had made. They noticed that even their lower production quality videos performed well when they were responding to what the viewers had asked for. But in crunching numbers, they also realized that their subscriber conversions were a lot lower than other channels, so they started experimenting with different types of content to see what would work best. Eventually, they embraced that their audience was getting younger and younger, so they decided to make content geared toward 6- to 12-year-olds. It worked. Their channel blew up. It worked because they kept trying to understand their audience and

changed their content to match. It's important to reiterate here that
you have to create from a place of passion, so don't change your content to match at the expense of doing something you love!

Find the Sweet Spot

One of my favorite examples of someone who found their audience is
Devin Stone. Devin is a lawyer who started making YouTube videos
to help law students prepare for the LSAT and survive law school. He
found a good audience on his channel *Legal Eagle*, but he wanted his
channel to grow. Devin reached out to me, and I had a consultation
with him to give him tips and advice to grow his channel. I asked
Devin how many law students there were in the United States at any
given time, and he said around a hundred thousand, so I helped him
understand that he had reached the threshold of market penetration
with what he was doing, and if he wanted to grow past that, he had to
extend his reach.

Devin's audience was maxed out with the kind of content that
he was creating, and if he wanted to get a bigger audience, he had to
change his content to reach people outside of the law student demographic. I asked him to reiterate why he wanted to make videos in
the first place, and he said he just wanted to share his passion for the
law and help people understand law better. So the follow-up question
was: How can you help more people outside of law school understand
the law? The answer we came up with was to use law-related TV
shows and movies a mainstream audience already enjoyed watching
and deconstruct them in a "Real Lawyer Reacts to . . ." format. He
already had one video in this format, but he needed to create more in
this series. He also created an original true crime special. He was still
reaching his original viewer, but he had found a way to appeal to a
pop culture viewer as well. *Legal Eagle* now has more than a million
subscribers and a hundred million views. Devin figured out how to

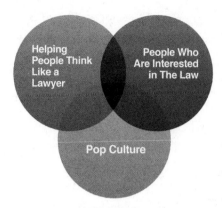

Figure 11.2 Scalable valuable content

cross his passion with his audience, but the real sweet spot came when he added the third dimension: the mainstream/pop culture. Take a look at the Venn diagram in Figure 11.2 to see exactly what I mean.

Let me stop you right here, because I know what many of you are thinking: "But my content is for everybody." No, it's not. If it is, then that's your problem. You have to find your audience first. Recognize who your "love" group is for your niche, and expand to a broader audience from there. By "love" group, I mean your loyal fans. These are your groupies. Think of The Grateful Dead's fans; they're so loyal, they even have their own groupie name, "Deadheads." These are the people who go to every concert—even if it's the same concert—in every city, taking time away from work, relationships, and hobbies just to follow the band. Your online groupies, when you find them, will consume whatever you make just because *you* made it. They've been converted to you.

Matthew Patrick, known on YouTube as "MatPat," made a video in 2020 to talk about some pretty deep issues for creators trying to make a living on YouTube. MatPat's channel, *Game Theorists* is obviously for gamers, so in order for this out-of-the-ordinary video to perform well with his "groupies," he brought in an Infinity Gauntlet analogy to explain the issues in a way they could understand and want to keep watching. But do you know who else is familiar with the Avengers movies and the Infinity Gauntlet? The mainstream. MatPat nailed this video, both for his loyal fans and for a broader audience, and kept them watching because they liked *him,* and because he used something widely relatable.

Not All Channels Are Created Equal

Not all channels work the same in the Venn model. You will see different results based on the type of channel you have; that's why it is so important to know who your love group is before extending to the mainstream. If you have an educational channel, you shouldn't take what a gaming channel is doing and copy it exactly then wonder why their successful strategy didn't work when you tried it. Content resonates differently with different audiences. Someone who likes to watch gaming videos is going to have a specific viewing behavior toward gaming content and a specific viewing behavior toward, say, a cooking tutorial. And vice versa.

There are a bunch of different types of content: how-to, best of, vlogs, product reviews, cooking, gaming, music, pranks, kids educational, and the list goes on and on. It should make sense that if your channel is a vlog that your viewer is going to have different patterns from a how-to viewer. Ultimately, people consume content they care about, so figure out who cares about your particular type of content.

Be careful to assume, though, that this means a food viewer is a food viewer on all content about food, for example. You really have to narrow it down beyond a broad category to understand your ideal viewer. In other words, someone who wants to find and follow a recipe doesn't want you to add your commentary, personality, and visual creativity; they just want the recipe. Whereas, another channel might be perfect for that viewer who does want to be entertained and connected to the person making the food. See how specific you can get? When you can identify who is watching and where and what they enjoy, you've hit the jackpot. Think of Gordon Ramsay. People who watch his content are not there for the recipe, they are there for the entertainment. The food is just another character in his story; it's not what you're going to make for dinner tonight.

The AI Needs to Know Your Audience, Too

When you can get the YouTube algorithms to understand who your viewers are, they will do the work to find other viewers with similar viewing patterns to serve your content to. Remember that YouTube's #1 goal is to keep the viewer on the platform longer, so if you've honed your content to match a specific viewer, YouTube will help your content be seen by that viewer who will likely want to watch it. The AI puts similar videos together to be recommended with like-minded viewers because, well, it works. The faster you can make content to fit in a collection of videos that will be batched together, the faster it gets pushed in recommendations and discovered by the right viewers. To stand out among them, though, use strategies to get high Watch time and AVD. I show you how to do this in Chapter 16.

I truly believe there is an audience for anything you could possibly want to do on YouTube. Canada's self-proclaimed "saltiest" creator, Jackie NerdECrafter, found her audience when she stopped trying to do it the traditional way and started being herself. "I thought I had to try to be pretty so I can attract people because I am too plain," she said. "So I started wearing eyeliner, but I was allergic to it and was tearing up in my videos. When I realized that I was going to get negative comments even with eyeliner on, I ditched it."

Jackie embraced her authenticity and showed her personality and her mistakes, and that's when the growth occurred. She attributes her growth to being herself; in fact, she thinks this is why her channel took off compared to other female channels that came before hers. They were trying to be perfect; she was being genuine.

Jackie said, "When I posted my first video on YouTube, it got 30 views. I thought, 'Holy moly, 30 views; that's amazing! Who are these 30 people?'" She really cared who those people were, and she wanted to understand them on a level where she could continue to make content they would watch and like. This should be your goal whether you have 30 viewers or 30,000. Ask yourself who they are and what they might want to watch.

But . . . *How* Do I Learn about My Audience?

This might sound crazy or impossible, but I like to figure out who my audience is before I figure out my niche. That's how important finding the right viewer is. Don't get ahead of yourself with creating content, because, if you do it right, your viewers' behavior will change your content. This is where we could talk about the whole cart-before-the-horse analogy. People often create content fast without fully understanding who they're creating for, which is a great recipe for frustration. I like great recipes, but not this one. Learn who you are creating for. Your content will be naturally consumed and shared and won't lead to frustration, so take a step back and get to know your audience first.

The Persona Breakdown

In order to accomplish this, I do some educated guessing before I have the data I need. I call it a Persona Breakdown. The persona is your viewer, your avatar, your target audience. You want to get to know that person as well as you can. This is tricky to do right out of the gate, but you can make assumptions until you have the data to help you.

Figure 11.3 Persona Breakdown

Take a look at the Persona Breakdown shown in Figure 11.3. If you want to get to know your viewer so you can make content they'll watch, you have to know things about them demographically, psychographically, and behaviorally, both online and offline. I go through this breakdown twice: once for a male viewer and once for a female viewer.

First, demographics. Think about age and gender before anything else, then consider income range, education, location, relationship status, and children status of your viewers. Don't you think it would be helpful to know if your audience is a single male majority between the ages of 25 and 35? If they're girls? Gen Z? Middle class? I know some creators who assumed they were making content for teenage girls, but when they had a live event, the overwhelming majority of attendees were actually eight to nine years old. This fact surprised them, but they changed their content to really cater to that younger audience, and their channel performed exponentially better.

After demographics, look at your viewer's psychographics. Demographics are external traits or facts about a person, while psychographics are their internal traits. Psychographics break down a person's beliefs, values, attitudes, motivators, lifestyle choices, fears, and vulnerabilities. It's what drives them. It's their goals and aspirations; it's what they are passionate about. Demographics are the boring facts;

psychographics are where it gets fun, because you get to find out who your viewer really is. They're no longer a number; they're a person. You can now connect to them on a level you couldn't do otherwise.

A great example of seeing the person behind the number is when a celebrity connects with a fan one-on-one. You've seen the videos. Once of my recent favorites includes Billie Eilish sending a personalized video to a 13-year-old leukemia patient. Billie solidified a fan for life, not only in the girl who was sick, but also in people like me whose hearts were touched deeply. Think about how that gesture affected Billie's "love" group and their loyalty to her.

After looking at demographics and psychographics, you might have a pretty good guess about who your viewer is, but none of that matters if you don't know what they actually do. You have to know how they act out their daily life, both in the real world and online. YouTube's algorithm can't think about your viewer's psychographics like you can, but what it can do really well is observe your viewer's online behavior. In fact, it sees data we can't see as humans. Even with more than two billion active, logged in viewers, it can still watch every viewer to know their behaviors. Specifically, the AI watches what a viewer does and doesn't do when they click on a video. It observes what a viewer searches, clicks on, doesn't click on, watches for five seconds and leaves, watches for the duration of the video, clicks on next, and so on. The AI connects the unstructured data and similarities between viewers.

Your Viewer Lives In Micro-Moments

You need to know even more about your viewer's minute-by-minute daily behaviors, though. These are what Google calls "micro-moments." Google put together an excellent guide on how people interact with online content and how it affects their real life. The gist is that people always have their mobile device near them,

they check it 150 times a day, and the majority of time spent on
it is made up in moments. These moments are defined by intent:
they want to know something, go somewhere, do something, or buy
something. A brand needs to be there to answer any of these queries
quickly, because the opportunity to put content in front of them in a
useful way will only last a moment.

Here is a quick example of each type of micro-moment:

- **The I-want-to-know moment:** You hop online and do a search
  for a random fact. This happens a lot in everyday conversations.
  A question comes up, and somebody says, "Google it." So you
  hop online to get the answer, and hop off.

- **The I-want-to-go moment:** Probably the most common one
  here is something like the "find Italian restaurants near me"
  type of search. You want to go somewhere nearby for something specific.

- **The I-want-to-do moment:** You need to change the air filter in
  your vehicle, so you search how to do that on YouTube.

- **The I-want-to-buy moment:** Your toaster fried itself this morning and you need a new one ordered today. You are ready to buy.

Think about your own mobile habits from day to day. Do you
take out your phone to google something and spend the next two
hours casually browsing through an entire website, or do you take
out your phone when you have a thought pop into your head about
something, click on the first result that looks the most relevant, and
ascertain within a handful of seconds whether you will stay on that
site? When you find what you're looking for, you put the phone away
and go back to whatever task you were on. Five minutes later, you
pull out your phone, text your friend, check your social notifications,
and put your phone away. Fifteen minutes later, the phone comes out
again, tells you the closest Mexican restaurant with good reviews, and

goes back in your pocket. At the restaurant, you open your phone to take a picture at lunch, send it or post it, and put the phone away. We live in micro-moments, online and offline.

We all live in this binary, but we do have differing habits. In planning a trip to Europe, for example, some of us plan months ahead, researching the best places to visit and eat, reserving hotels and tours in advance. Others book a last-minute trip when they see a bargain, and they navigate the details on the fly. My daughter Ellie is this way. She came with me on a business trip to Berlin, and I asked her to figure out where we were staying and what we would do. After my speaking gig was over, I asked her what the plan was, and she pulled out her phone right then and found us the best burger place hidden away in a converted subway bathroom. And the food wasn't crappy; it was delicious. Her online behavior decided what we did in real time, and where we went afterward. If my wife Carolyn had been the one along on the trip, she would have planned the details of our trip months before we packed our bags.

Putting the Persona Breakdown to Use

The Persona Breakdown applies deeply to how you create content. Know your avatar's details and habits so you can give them exactly what they want.

Let's do a Persona Breakdown with a hypothetical example. Remember when I said you can't lump all "food" viewers together? The Gordon Ramsay viewers are not the same as the I-want-a-recipe viewers. So let's say we want to make content for a food audience, but we want to target "foodies." Foodies are a completely different group of people than the recipe type. Who would fit the foodie mold? I'm just going to make some assumptions here. Demographically, I'm going to assume they are between the ages of 25 and 40, skewed majority female, income range between $40,000 and $120,000, most living in bigger metro cities, with some form of education after high school. Locationally, I'm going to assume a lot of them live in one of the three top states of YouTube viewers: California, Texas, then New York.

Next, we'll think about this viewer's psychographics: these are people who aren't necessarily cooks but they enjoy food socially. They like to take pictures of their food and post to Instagram. They like to recommend great places to eat to their friends. They like to be entertained, not just fed. They are often the working class nine-to-fivers, and this kind of activity or hobby is something that brings them pleasure.

As far as their online consumption, they gravitate toward pretty thumbnails and good entertainment. They might subscribe to Gordon Ramsay and similar channels. They watch cooking shows on Netflix like *Cake Boss* and *British Bake Off.* They're into "food porn."

Their offline behavior mirrors millennial habits: they might take time off from their very stable job to go do what they want; they might take an extended vacation; they're less apt to get married and have kids young.

Now here's the golden zone: the merging between their online and offline behaviors. They take a photo of their food and post it on social media. They eat at a restaurant then review it on Yelp. They love to find the hidden local pizza place and tell their friends about it online. When they're at a party, they talk about places they've been and make decisions on where they'll go in the future based on those conversations. They're more opinionated. They're more likely to watch *Cake Boss* to agree or disagree with the judges.

Remember that we're just guessing here. You can make a more educated guess by doing some Google searches about your topic or the people you think would fit into your audience. The more content you create and observe patterns, the better you'll get at guessing. And

then, as we get more data, we can decide what to do based on that
info because the data isn't a guessing game. You'll be able to see all
the demographics and online behaviors in your video analytics and in
their comments. When you have the analytics in play, then you can
go and gather more information and take it to the next step, collect-
ing outside information about your viewers' online behaviors. The
more you know, the better you can cater to them.

These Are My People

By evaluating our viewer in a Persona Breakdown, we loosely defined
what we are assuming the viewer is. Next, we get to do our research to
either validate or dismiss the assumptions we just made. We will cover
how to hone in on your viewer by doing recon and research in the
next chapter. This is the geeky fun part I'm always excited to share;
most creators have no idea what they've been missing in their content
strategy, and they are blown away by insights they find. I promise
you will figure out things about your content and your audience that
will change your strategies and help you create better content. You'll
learn how important it is to make a plan—it's the foundation of
your success.

There are so many different types of people in the world. Who
they are and how they behave online and offline depends on how
they were raised, their social status, their lifestyles, their personalities, and more. When you make videos in the right way, it's a magnet
for certain types of people, who will then consume that content in
the same way another viewer with similar traits and interests does.
YouTube finds your like-minded people and gives them your content
to consume.

There are two ways to look at your avatar when it comes to You-
Tube: as someone who will view your content, and as someone who
will share your content or potentially buy something. Some disagree

that these are two different personas, but I believe it's a valuable distinction to make when it comes down to who your viewer is and what their intention is with your content.

My grandma, God rest her soul, was a buyer persona. When she passed away, we had to sort through tens of thousands of dollars worth of things she had purchased on QVC . . . and never used. A good advertisement always got her to pick up the phone, credit card at the ready. For better or for worse, I'm the same way: when I see an ad that shows me how something can functionally improve my life, even if I never would have thought I "needed" that item before, I buy it immediately. My basement is starting to look a lot like Grandma Eves's QVC haul.

In the next chapters, we go in depth on doing research and making good content so it gets easier to find your audience, but to start, the Persona Breakdown is a great way to get a feel for your potential viewer. Then as you make data-driven decisions based on your viewer's traits and behavior, the magic will happen. I love using the data to make decisions because it's no longer a guess at who your viewer is and what they are doing: the facts are right there; the data doesn't lie. You'll finally be able to say with confidence, "These are my people."

Action Exercise

Create your viewer Persona Breakdown. If you have channel content, check your analytics for your viewers' information to complete these tasks. If you don't have content, you can still do this as a projected Persona Breakdown until you have data to pull from. Do this twice, once for a male viewer and once for a female viewer.

Task 1: List your viewer persona's demographics: age or generation, gender, income range, education, geographical location, and relationship status.

Task 2: List your viewer persona's psychographics: motivators, values, attitudes, lifestyle, fears, and goals.

Task 3: List your viewer persona's online behavior: types of media consumed, content consumed for personal interest versus for entertainment, and channels they subscribe to.

Task 4: List your viewer persona's offline behavior: buying behavior, habits, hobbies, and where they spend their time offline.

Get the companion workbook and find more resources at www.ytformulabook.com.

12 Recon and Research

Every business knows there is no business without the customers. You have to get the people in the door. This is called acquisition. A lot of money and time is spent on marketing to acquire customers. You need to spend some time figuring out how to get the people in the door, too. You might not be seeing your viewers as customers or consumers, but you should. They are consumers you need to acquire to be successful. I'll show you how to find your people, and more importantly, how to retain your people. You have to know how to keep them coming back for more, or you will lose them. Our goal is to keep the loss or "churn" rate as low as possible and the growth rate continuously climbing. Every good business does this well, and so should you.

This is going to be one of the funnest chapters of the book because we get to talk about exactly how to go and find your viewer tactically. The funnest part is that it works.

In this book's introduction, I said that a pooping unicorn made me write this book. This wasn't tongue-in-cheek; I literally meant it. Here is the story behind it: I was visiting my friend Jeffrey Harmon, whom I have worked with on many projects over the years. We complement each other really well professionally. He told me about a potential new project he might be working on that would have a pooping unicorn in its ad. He had me at unicorn, but he really had me at poop. The ad was for a company called Squatty Potty, and they had a challenge to overcome. Their demographic was the boomer generation, and they had plateaued. They knew they needed to reach a younger, health-conscious avatar, and they reached out to the Harmon brothers' marketing agency to help.

I told Jeffrey I really wanted to be a part of the project (Who wouldn't? An ice-cream pooping unicorn? Sign me up!), so we joined forces and started on the creative. We needed to take the "ickiness" out of going number two, and we had to convince the Squatty Potty owners and their *Shark Tank* investor that this would work. Originally, the plan was to use an artificial Clydesdale-sized unicorn to poop because it was big enough to fit an ice-cream machine inside, but Daniel Harmon, Jeffrey's brother and business partner, had the idea to use CG effects and a small unicorn instead.

We needed to know where to find the unicorn lovers of the world and figure out who they were on an individual level so we would know how to connect with them in our message and delivery. Every marketing campaign's success hinges on knowing who the buyer is, but I wanted to know more than who would buy a Squatty Potty; I wanted to know what kind of person cared about unicorns enough to respond strongly to an advertisement that featured one. So I went on Reddit, which is always a great place to go and find niche groups, and I was looking for people who liked My Little Pony, because we thought this would be a community that would respond to the unicorn.

There is a subreddit for every topic, and wouldn't you know, I found a community of My Little Pony fans . . . who were grown men. They even called themselves "Bronies," and they had their own culture of My Little Pony memes, terminology, and costumes. I'll never forget the costumes. I lost three days of my life to the Bronies, because I just couldn't believe such a bizarre subculture existed, and I kept getting deeper and deeper into the rabbit hole. It fascinated me so much. I went back to Jeffrey and Daniel Harmon and Dave Vance, the head writer for the project, and told them about what I had found. We kept the Bronies in mind as we brainstormed and created the ad, because we wanted to make something that would resonate with their subculture—we wanted them to like it enough that they would talk about it and share it with people. Dave came up with the magical world and the prince, and wrote a brilliant, entertaining script.

The Brony discovery moment was monumental for me in my work because it made me realize just how important it was to go find your avatar. The lightbulb had gone off; it was a very aha moment for me. It's important to note that the Bronies were not our target buyers. They were the focus group who would respond to the video ad naturally. They would create a feedback loop for the campaign: they would find the ad and watch it all the way through, comment on it, and share it. They would create the buzz so those viewers that would buy the product would see the ad. They would kick-start the campaign and get it to grow.

For every 10,000 views, we knew how many shares we would get and how many people would tag their friends, engage, or purchase. I handled the organic pushes to different viewer personas that would view, consume, share, and buy. Jeffrey looked at it from the long-term approach of acquisition and retention. He measured the buyer persona by one criteria: who swiped the credit card. Between us, we had all the bases covered. Like I said, we work great together.

When the ad was ready to launch, we knew we had something big. It was a longer form video, almost four full minutes, which was much longer than traditional advertising, but it was original and extremely entertaining. Jeffrey and I had a launch strategy session, and we had the idea to make a gif of the unicorn pooping ice cream with the caption, "The world is going to change on [this date]." What did we do with it? Leaked it to the Bronies, of course! I privately shared the animated gif with a member of the Brony community. How does the saying go . . . You can lead a Brony to water but you can't make it drink? Well, they drank. They were wild about the gif, and when the video was released later that week, it hit the front page of Reddit.

We got more than 20 million views on the video in less than 24 hours, and it kept gaining steam every hour. Jeffrey and I were dumbfounded at the response. The best part was that we knew that for every dollar spent we would make a certain amount. We put fuel on the fire by running ads against it on top of its organic distribution. It has sustained sales over time, which is great because we were able to analyze and adjust along the way, and it has kept working.

In the ad's first year, it brought in $28 million in attributable sales, and to this day, years later, it's converting every time the ad runs. There are still people who haven't seen the ad, even though it's several years old, and we know they will buy because we know the persona, so we can target exactly who we need to. There are new buyers coming on the market every day, and when the right one matches, we've acquired another customer.

I haven't seen an ad go viral like this one before or since. It was even dubbed, "The greatest viral ad in Internet history," by *Boing Boing*. This is how I found the true secret of understanding content creation based on finding the right person for the content. We found the right viewers who would share the ad, and we found the right viewers who would buy the product. This is the secret magic of the ice-cream pooping unicorn.

Reconnaissance

Reconnaissance means you are in discovery mode. You need to discover your own Bronies, so to speak. Your job here is to nail down what type of content you want to make. You want to know if anyone is doing something similar on YouTube already, and if so, what each channel stands for and what their audience cares about. You want to know who the big creators are in the niche—who has movement and momentum. You want to figure out who the target audience is and how they are responding.

So let's get down to business on the *how*: Go to YouTube and make a list of 20 different channels in the niche that you are interested in. Then go to the first channel's videos and sort by most popular. Watch the most popular videos that were uploaded within the past year. Go to the next channel and do the same thing. You need to write all of this down. Pay attention to what videos are recommended next when you're watching each video. You'll be able to see what is working on YouTube right now. Some of these channels might have a lot of subscribers, but they haven't seen recent activity. Make sure the channels you are listing have active viewers who have recently consumed the content and/or commented. Some of these big channels haven't posted a video in a really long time, so don't use these ones for your recon purposes. When you sort by most popular, make sure they have something recent in their top results. Look at how engaged the viewers are on each video and channel. Look at the viewer-to-subscriber ratio to see how big or small the market size is for the niche. Remember Devin Stone from Chapter 11 whose *Legal Eagle* channel teaches people about the law? While you're doing recon, keep in mind how Devin reached out to a broader audience than the law student crowd.

What you are doing is gathering enough data on similar content to the type you want to be creating. Grouping similar channels together helps you see patterns. When you understand similar channels and how their viewers interact with and respond to their content, you are figuring out who your viewer might be.

I don't want you to do deep research on these channels just yet. Your purpose is to grab as much information as you need to discover what's out there in your niche and who those viewers would be. You'll analyze all of this information in the research phase later.

After you make your list on YouTube, go search the niche off of YouTube as well. The people who would be interested in your type of content are interested in things elsewhere on the Internet, and you need to understand their behaviors and interests. Reddit is a great place to find communities. You will be surprised about the things your viewers are interested in that you didn't even know existed. Once you have all of this information, you're gearing up to make data-driven decisions.

Great creators are always in reconnaissance mode, even after reaching huge success. You don't do recon once and call it good. There is always more to discover about your audience, and your analytics will continuously show you patterns. You will discover hidden gems when you diligently watch your data. Look at the data with the question, "Why?" always in your mind. Why did your content succeed or not succeed. Why did your viewers interact with your content the way they did. But be careful not to get stuck here: a big mistake I've seen creators make is spending too much time in recon mode. It's so easy to fall into the rabbit hole when you start digging (read: my journey into the annals of the Brony world). Get the information you need, and move on to the research phase.

Research

Okay, now you are ready to dig in and do some analysis on the stuff you gathered in recon. Go back to your list of YouTube channels and click on one, then, again, sort the videos by most popular. Just forget about the older videos on the channel; you want to look at what has been working recently. Choose 6 to 10 videos per channel that have the most views and make notes on titles, thumbnails, video views, likes and dislikes, and video duration. Notice if the creator uses a hook to pull viewers into the content, what the pacing feels like, and how they edited. Ask what similarities you see among different videos and across different channels in your list. Make note of reengagement throughout the video, calls to action, video descriptions, and comments.

You are looking for patterns here: patterns in content creation and patterns in the audience's behavior. This practice is all about the response and interaction of the viewer with the content, and to see what is working right now in your niche. While this section of the chapter isn't long, the process of doing the research takes time. It can get tedious, but do not skip ahead. You have to do the work if you want to see the patterns. I promise it will be worth it; just stick with me.

The Ultimate Unicorn: Jesus

Recon and research can provide the perfect storm when it comes to finding your community and getting your content out to the world.

Another project Jeffrey Harmon and I are currently working on together is the polar opposite of a pooping unicorn; it's Jesus. I am more passionate about this project than any other I've ever done, and I've done *a lot* of projects. We wanted to create a TV series about the life of Jesus Christ that appealed to the evangelical community and had the professionalism of an HBO-quality series. Jeffrey introduced me to a creator named Dallas Jenkins who later became my partner and the writer and director of the series about Jesus Christ called *The Chosen*. Ricky Ray Butler, Matthew Faraci, and Earl Seals also joined as owners. Jeffrey wasn't going to be an owner, but he was going to distribute the product with his VidAngel company.

On a big project like this, we do a multiday lockdown marketing session, but beforehand, we have several brainstorming sessions to get a good handle on who the right viewer persona would be: who would be passionate about it, and who would back the project. I thought it would be easy to do because it would be people just like us. Dallas knew we could speak the message but was more skeptical that we could get people to invest. We had to meet the needs of our project through crowdfunding, and he said he would be impressed if we could make $800.

Matthew had worked on several big evangelical projects and knew the space well. Dallas also had a lot of experience in the space. But no one had ever made HBO-quality Christian productions. Texas is one of the largest groups in online video consumerism, so we targeted Texas and other southern Christian areas. In other data that we had gathered in recon and research, we realized that Gen X and millennial women are the biggest spenders online, so our target buyer persona was females aged 25 to 45. We targeted the people who were the community, school, and church volunteers, the I-love-Jesus type.

For our lockdown marketing session, we got ourselves an Airbnb, ordered in food, and got down to business. We laser-focused on who our audience was. We defined the path of how we would get from where we were to our end goal. We had to figure out how we would raise the money and how we would find the army of people who would resonate with our content and promote it to the world.

We had seed money with the intent of crowdfunding to reach the needs of this project. At the time, *Mystery Science Theater 3,000* and *Veronica Mars* were the highest grossing crowdfunded projects with devout, cult followings. So we made a goal to surpass these projects and become the #1 crowdfunded project in the history of movies and television. The goal was $10 million. We knew we wanted to resonate our message and vision with a specific persona (25- to 45-year-old female, married or unmarried, church-goer, volunteer). Then we asked where do these people congregate online? The answer was not on YouTube. We defined that Facebook and Instagram is where we would find them. We developed our messaging and content strategies around this precise audience and targeted them primarily on Facebook, because it was the easiest platform to share the content and the message of our campaign with like-minded people.

We launched the pilot episode, which told the nativity story about the birth of Jesus, and we pushed it to the right people at the right time: just before Christmas. The first episode hit the sweet spot in the Venn diagram (from Chapter 11) because it was timely, topical, and could reach our ideal viewer in addition to a broader, pop culture audience.

We more than doubled the all-time crowdfunding record with a show about Jesus. You might even say it was like manna raining down from heaven, if you're the Bible type. Some of our biggest contributors said they donated money because someone had pushed it to them—that someone was usually a person who fit our target persona profile (our buyer strategy worked!). In just a few short weeks, we were able to get tens of thousands of Facebook followers.

I knew that in order for us to massively grow organically, we needed to nurture our audience on other platforms. It took me two years to convince Dallas to fully embrace becoming the "influencer" face of our project, and that's when everything changed. People identified with him and his passion, personality, and authenticity. When we went live on *The Chosen* YouTube channel, Dallas started speaking to our avatar, YouTube found the viewers who matched, and suggested our content to them. We got 146,886 subscribers in 14 days. Yes, 14 days, and yes, that exact number. Every single one mattered to us.

We can't get more opposite in demographics and content than in the two examples in this chapter, which I actually love because it demonstrates that the method works, no matter what your content is or who your audience is. Speak to your target audience, and then broaden to get your content to the masses. Upcoming chapters will show you how to really leverage that.

The essence here is that the more you understand and relate to your audience and create content for them, the more YouTube will connect the dots and feed them their preferred flavor of ice cream, so to speak. Maybe it's the unicorn-poop-flavored kind, or maybe it's the Jesus kind. Our recon and research made it possible for us to form a community of people who loved and distributed our message and content. This is what can happen when you do enough recon and research; this is the payoff. This is why you do everything in this book.

One word of caution here: this isn't a one-and-done step. You don't do recon and research at the outset and cross if off your to-do list. You have to keep going back to look at your data and reevaluate. The more data that comes in, the more patterns you will see and the better you will be able to shift your strategy as needed to make better decisions about your content.

Action Exercise

Perform the reconnaissance steps to better find your audience and see what they're consuming on YouTube:

Task 1: Research at least 10 successful creators in your niche.

Task 2: Take note of what they have in common. Watch their most successful videos and take note of video creation and editing patterns.

Task 3: See if and how they engage with their communities, both in the video comments and on the Community tab. Read the comments on their most popular videos.

Get the companion workbook and find more resources at www.ytformulabook.com.

13 Content Is King

May 26, 1980.

That date will always be seared in my mind because something changed me that day, and I haven't been the same since.

This chapter is about moments that change people. It's what I am the most passionate about in my work—figuring out how our message or "content" can make the biggest impact on others. Some of the greatest experts in rhetoric have changed the world, both good and bad, from Aristotle to Hitler to Gandhi, with their messaging. Every interaction we have in every facet of our lives comes down to getting a message across. Messaging is a huge part of our YouTube strategy. It helps us get people to do what we want them to do and meet our success goals.

On May 26, 1980, I was six years old. My two uncles had gone to see a new movie that had opened a few days earlier, and they loved it so much that they needed a reason to spend money to see it again. So they asked if they could take their kid nephew Derral. What a great idea! We got to the theater and found our seats, and the title sequence began its trademark Star Wars crawl up the screen. It was *The Empire Strikes Back*—the second film in the original Star Wars trilogy. Now, I don't say this lightly, but *The Empire Strikes Back* is

THE BEST MOVIE OF ALL TIME THAT EVER EXISTED IN THE WHOLE OF HUMANITY. This was content at its finest . . . I knew it even at the age of six. The cinematography and the story were like magic to me. But there was that one epic moment in the movie, you all know it now, when Darth Vader revealed that he was Luke Skywalker's father (oh, and Luke's hand got cut off, too, by the way). At the time of filming, only a few people even knew that scene would be in the script: Mark Hamill (Luke Skywalker), James Earl Jones (voice of Darth Vader), film creator George Lucas, and the director, Irvin Kershner. Hamill says he didn't even know about the plot twist until right before filming the scene, and Kershner threatened him with a "We'll know it's you" if the twist got leaked.

The shocking sequence branded itself into my brain. This moment changed my life. I became obsessed with everything Star Wars (even the cringy content, like Luke's passionate kiss with his sister, Leia, and the *Star Wars Holiday Special*) including all the movies, posters, and other memorabilia. I coveted all the collectibles that a poor kid with nine siblings couldn't buy. My parents bought *The Empire Strikes Back* on VHS, and I have watched the movie 16,482 times since then (no longer on VHS, though). (For the kids: a VHS gets inserted into a VCR and plays a movie, much like a DVD to a DVD player. If you don't know what a DVD player is . . . I can't help you.)

Ads with Impact

Fast-forward a few years. I was watching TV in 1984 when a commercial came on for a Wendy's hamburger. I was already an ad nerd, even at 10 years old, but this ad struck me. There were three old ladies looking at a big bun talking about how big it was, and when they opened the bun, there was a tiny burger patty inside. One of the old ladies then says the classic ad line, "Where's the beef?!" I laughed and laughed. This was great marketing! I told my brothers and parents and everyone I knew about the ad because of its impact on me; it was so funny and different. Of course, we didn't have DVR or the Internet, so I waited by the TV with my finger on the VCR's record button so I could show everybody what I was talking about. It was the 1980s way of hitting the share button or tagging someone on a post.

Later that year, the Dunkin' Donuts ad "Time to Make the Donuts" had a similar effect on me. In the ad, Fred the Baker went in and out of a door saying the catch phrase "Time to Make the Donuts" each time, and at the end of the ad, he catches himself off guard by going in and coming out of the door at the same time. It was like time had folded onto itself, and he was seeing double. I rolled. What a fantastic surprise! At this time in my life, I was consumed by ads like this. I wanted to know why they worked. I wanted to learn how to create content equally brilliant and effective in its messaging and impact.

In 1987, PSA ads were my teenage obsession. They were so striking and powerful in a disruptive, original way that hadn't been done before. Ads like, "This is your brain on drugs," and, "Didn't see that coming?" had a shock value that worked. In the first ad, a guy takes an egg and says, "This is your brain." Then he shows a hot frying pan and says, "This is drugs." Then he breaks open the shell over the frying pan, saying, "This is your brain on drugs. . ... Any questions?" as viewers watch the egg get fried. I was blown away by the power the message produced in such a quick and simple video. The ad was less than 30 seconds long, and I will never forget it. Of course, YouTube wasn't around when this ad aired, but it's fun to go to the replay on YouTube today and find comments like, "That's my brain while doing math. Not meth . . . MATH."

The "Didn't see that coming?" PSA ads from the US Department of Transportation had a similar shock-value effect. In one video, it begins like it's advertising a minivan, listing all the specs and selling points while a family loads up after the son's football game. Suddenly, another vehicle slams into the side of the minivan where the son was sitting. The screen goes black and says, "Didn't see that coming? No one ever does. Buckle up. . . ." This was extremely powerful advertising; the message was loud and clear.

I also started watching *Saturday Night Live* (*SNL*) skits in the 1980s and 1990s that were so funny and so memorable to me. I recorded the skits on VHS tapes so I could watch them on repeat and share with my family and friends. On one occasion when I needed to record a show, I accidentally grabbed a VHS that was sitting near the TV that had my brother Joe's school project recorded on it. It was a cute little video of Joe demonstrating how to milk a cow . . . until I recorded over the top of it. What content had been so important to capture that it was worth sacrificing precious family memories for? It was a sketch called "Colon Blow," a fiber cereal spoof that ends with the warning "May cause abdominal distention." My family will never let me live that one down. Other favorites were "The Love Toilet," a combo-john for obnoxiously attached lovers, and "Oops! I Crapped My Pants: Undergarments for the Elderly," which is pretty self-explanatory. Looking back, I like to think of these ads as precursors to my work on Squatty Potty's ice cream–pooping unicorn ad. What can I say, my penchant for potty humor runs deep.

Of course, the much anticipated event in advertising came around every February when it was time for the Super Bowl. I anxiously awaited the Super Bowl for the ads as much as any football fan did for the actual game. I recorded every commercial during the game. In 1984, Ridley Scott's Apple commercial introduced the world to Macintosh computers during the third quarter of Super Bowl XVIII. In 1949, novelist George Orwell had written *1984*, a book that portrayed a dystopian society under the thumb of a corrupt government. Since it was now the year 1984, Scott created his own version of the story—an ad boldly proclaiming that its product would free the people. The ad featured an athletic female running toward a giant TV screen, which was broadcasting a message to a throng of brainwashed people. She had a sledgehammer in hand, which she hurled toward the giant screen, shattering it just before the authorities ran her down. The message then says, "On January 24th, Apple Computer will release Macintosh. And you'll see why 1984 won't be like *1984*." It was such a confident claim—and such brilliant messaging. (Interesting fact: *1984* author George Orwell's estate and the novel's TV rightsholder sent a cease-and-desist letter to Apple and their ad agency, claiming copyright infringement. The ad aired only once nationally but won awards and was dubbed the TV commercial of the decade.)

I wanted to figure out the elements in epic ads like this so I could make my own. I wanted to be the guy who came up with the idea for the Super Bowl ad that people talked about afterward. I literally conducted surveys with my family and friends to see which ads were their favorites and why. Eventually, I went to school for marketing and advertising, which was a surprise to nobody.

Like these ads and the *SNL* skits, all good videos follow a pattern: they grab your attention with a hook, hold your attention with reengagement strategies, and leave you with a payoff or unexpected surprise at the end, like the great Darth Vader revelation in *The Empire Strikes Back*. In the ad survey–taking days of my youth, the results of every survey reflected that the clear winners were the videos that followed a specific pattern of narration to keep viewers engaged and had a memorable payoff at the end.

What Is Content, Exactly?

The phrase "Content is king" has been used before, but it is the best way to emphasize how true the statement is. As you can see from my own career path beginning at age six, content can change people's lives. Its messaging inspires, educates, shocks, and changes people to the core. When we talk about "content" on YouTube, one of the biggest mistakes people make is thinking that "content" equals "video." Video is definitely a part of it, but content also includes the title, thumbnail, and other metadata. Your video should be done well, sure, but you must give equal—if not more—attention to your video's title and thumbnail, because they are the make-or-break factors when a viewer is deciding whether they will click and watch. You will learn in Chapter 15 how to make a good title and thumbnail.

After you get them to click, you get to do the fun part: make the video! Scratch that: make a good video. You have to know what makes a good video in order to make a good video. It's simple circular logic, but it works. Like when Chris Farley says in the classic *SNL* skit, "You'll have plenty of time to live in a van down by the river when you're living in a van down by the river." We hear you, Farley. You'll make good videos when you know how to make good videos.

Pattern of Narration: Storytelling

All good content has good storytelling elements. People remember a good story even if they heard it decades ago. Narration in good storytelling and in good content follows the same pattern across the board, whether you're making videos about food or travel or education or pranks. And it works for kids, millennials, and boomers alike. It's called a story arc, and the best way for me to break it down for you is to refer to an episode of one of TV's most successful series, *Seinfeld*. The episode was titled "The Cafe," and aired in 1991. Figure 13.1 shows an overview of the story arc.

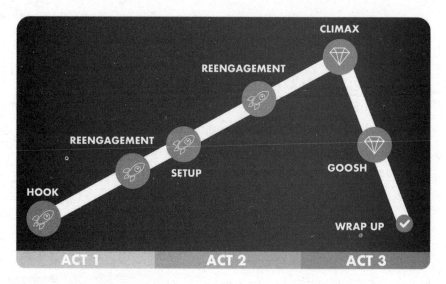

Figure 13.1 Story arc

The Hook

Comedian Jerry Seinfeld is a master storyteller, even though his show is commonly referred to as the "show about nothing." Most episodes for nine seasons began with Jerry performing a monologue standup routine on stage, telling jokes that foreshadowed the upcoming story in the episode. Its purpose was to hook the viewer right out of the gate and introduce the episode's topic. In "The Cafe" monologue, Jerry says that there's one shop in every neighborhood or town that is always changing hands: "It's a leather store then it's a yoga shop then it's a pet shop." He jokes that it's like the Bermuda Triangle—nobody can have a successful business there. This monologue is the "hook," element number one in our story arc. The hook creates enough curiosity for the viewer to want to know what the episode or video is going to be about.

Reengagement

Next, Jerry moves on to the reengagement element. The episode has begun, and Jerry says to George, "I haven't seen one person go into that restaurant since it opened. Poor guy." He and George have some back-and-forth. George complains about nice guys getting a bad rap, while Jerry theorizes that the restaurant owner probably has family in Pakistan waiting for him to send money. George weaves in two new storylines for the episode during this scene: the good guy predicament, and the girlfriend wanting him to wear cologne and take an IQ test. They've done a great job of foreshadowing and setting up these topics for further discussion and laughs in the episode.

Jerry is great at reengaging his audience so he doesn't lose viewers. In the episode, it shows him in his apartment watching the shop owner from his window with binoculars. He's encouraging passersby to go into the restaurant because he feels bad for Babu, the owner. He exhibits the "good guy" persona throughout the episode. Elaine (another of Jerry's friends on the show) asks Jerry if he's gone into the restaurant, and Jerry says he hasn't because he's afraid he'll end up partnering with the man because he's too nice. Then George walks in holding an SAT study guide, and Elaine sniffs him, to which George annoyingly responds, "I'm wearing some cologne, all right?!"

The Setup

The third element of the story arc is called the setup. It's the part of the video where you set up the coming climax, and it's another point of reengagement with the viewer. In Jerry's setup, he finally goes into the restaurant to support Babu. He thinks he's a nice guy by being a patron, and he even offers the owner some unsolicited advice to turn his multicultural, eclectic-cuisine restaurant into a Pakistani restaurant exclusively. Jerry gets excited about what's coming, because he thinks his "good guy" idea is going to be the reason for Babu's imminent success. This is a great setup. Be careful not to go straight for the climax at this point, because you have to keep teasing the viewer. This is where Jerry reengages the buildup by going back to his binoculars and patting himself on the back for helping Babu. As a viewer, we see that Babu has closed for renovation, and we want to see if Jerry's advice pans out in the upcoming climax.

The Climax

Now Jerry is ready to show us the climax of the story arc because he's built us up to it on several occasions. He goes back into the store after Babu reopens for business as a Pakistani restaurant. Babu is clearly upset at Jerry because he took his advice, spending time and money on renovation, and it didn't work. The place is empty. Jerry's "good guy" arrogance angers Babu, and the climax ends with Babu storming out of the room yelling at Jerry, "You're a bad man! You're a very, very bad man!" Jerry sits aghast that his good guy vibe has been crushed. The restaurant closes for business.

The Goosh

The story has concluded, but don't think it's over just yet if you want to retain those viewers. Really great content has a bonus element called the "goosh." I learned the goosh from Matt Meese, cocreator and head writer at BYUtv's sketch comedy show *Studio C*. This is the cherry on top, and *Studio C* does it extremely well. In the *Seinfeld* episode, Jerry is talking with his friends on the sidewalk in front of the closed restaurant. Elaine reveals her IQ score, making George feel bad, then Jerry asks what food everyone is up for. Jerry suggests Mexican, George wants Italian, and Elaine wants Chinese—an eclectic cuisine. Then Jerry says, "You know what would be great . . ." bringing it all back around to the original restaurant that had been there before Jerry opened his big mouth. George and Elaine glare at him, and the audience gets a last laugh. This is where you hide your hidden gem; it's the biggest value you have to offer, even on top of what the audience came for. Your viewers will come to expect this from you if you do it consistently, which is what will keep your retention up because your viewers won't dip knowing there is something still in it for them if they watch to the very end of the video.

The Marvel Cinematic Universe films are known for their goosh. When any Marvel movie ends and the credits come up, nobody leaves the theater. Why not? Because everyone knows there will be a bonus scene after the credits are done rolling. The Marvel bonus scene settles a plotline from the movie, offers comic relief, or hints at a future Marvel film. Whatever its purpose, though, moviegoers love it, and they leave the theater with that goosh moment in mind, which they'll probably tell their friends about. If you want to make a really great video, you gotta goosh.

The Wrap-Up

There is one last thing to tie up the story arc before the video ends: the wrap-up. On *Seinfeld*, almost every episode ends as it begins, with Jerry wrapping up the original joke he had introduced to the audience at the outset of the episode. In this example, Jerry talks about superheroes, the ultimate good guys, who hide their identity to keep people from criticizing them for the collateral damage that happens when they're saving the world. He's defending his "good deed" even though it didn't actually do anybody any good in the end. It's a fun way to wrap up the story in the last seconds that he has the viewers' attention.

If you've seen one episode of *Seinfeld*, you've basically seen them all. They are put together in this predictable pattern of narration because it works. Each story arc hooks the audience and reengages them throughout, weaving multiple storylines that converge in the end. There is a clear setup, an unpredictable climax, and a goosh for the bonus content or hidden gem. This pattern works for any genre, and if you follow it, you're setting yourself up for some royally great content.

Making Your Content Pop

My hope with this chapter is that you will never look at YouTube the same. As you watch videos, you'll train yourself to analyze every element, looking for patterns of narration. You'll learn how important storytelling is and how to emotionally connect with your audience. But I also want you to recognize all the little details in a video. The little details help you create content that really pops, details like camera angles and movement, time lapses and zooms, pattern interrupts, music, reactions, and editing techniques. Let me share with you an example of a creator who really knows how to make content pop.

I have worked with YouTuber MrBeast on a lot of YouTube videos and other projects, and I can tell you that his success is no accident. He is so thoughtful and methodical about every tiny detail in his content. He plans out all the elements of title, thumbnail, story arc, camera work, and every other little thing a creator could possibly think of when making a new video. MrBeast creates multiple title and thumbnail combinations, and films multiple options for different scenes in every video. His storyline is clear, his hook is bold and simple, and his reengagements are off the charts. He zooms, pans, and shoots from different perspectives. He edits with continuous reengagement in mind, doing quick cuts and fast paces because he knows that's what works with his audience.

He also keeps storylines simple. This content creation skill is highly overlooked and undervalued by most YouTube creators, but it's what sets great creators apart from the masses. If MrBeast can't explain a video concept in one sentence, he labels it as too complicated and scratches it. Some of these videos could have made him a lot of money, but he won't use it if it doesn't pass the one-sentence test. What he does do with these rejected videos is make them into a compilation video that he shows his reactions to. This way it's not wasted time, money, and content.

MrBeast knows how to make content pop. Case in point: he shot a video titled "I Opened a Free Car Dealership" (note the title less than 50 characters, easy to remember, and simple to explain), and the first thing he said was, "I bought every single car at a car dealership." This sentence flashes across the black screen in bold letters. We already know exactly what this video is going to be about: MrBeast bought a bunch of cars and he's going to give them away. Simple concept, masterful execution. MrBeast then shows himself buying the cars, which took a while in real life but about five seconds of video time. In this opening segment, camera angles switch multiple times, there's a quick time lapse, and there's comic relief including facial expressions. The content is already popping. Then MrBeast walks through the lot to take inventory of his newly purchased vehicles and changes the windshield pricing to absurdly low numbers. Hooked. This all happens in 45 seconds.

This video is nearly 17 minutes long, and you will want to watch it from start to finish if you go find it on YouTube. MrBeast is a wizard at reengagement strategies—he uses comic and emotional relief, camera work, and editing techniques to retain his audience and maximize his pattern of narration. MrBeast's retention timing is impeccable. He's listed the first car to sell for three dollars, but when the buyers sit down to sign the "paperwork" (which is a stack of crossword puzzles and word searches), he says, "We ran a 33% off sale, so it's now two dollars." All the while he's using zoom work, fast-paced editing, sound effects, and words on the screen to continuously reengage.

Authenticity and Emotional Connection

There is an important point to make here: this video seems to work because of sensationalism—MrBeast buys 12 cars and documents his silly antics to give them all away—but something happens outside of the sensation and the silliness. We see the happiness and gratitude of the people who receive the free cars, and we see MrBeast feeling really good about helping people. It's genuine, and MrBeast, it turns out, is a genuinely good guy. When your viewers emotionally connect, you have told a good story.

By the end of the car giveaway video, you'll find that you've smiled, laughed, and maybe even cried, so when MrBeast drops his call to action to buy his merchandise so he can continue spreading good in the world, you really want to click. Whether you'll click on the merchandise button or the next MrBeast video, he's got you. His masterful storytelling and content creation strategies worked like a charm.

If you grabbed a pen and paper and binge-watched MrBeast's channel with content creation techniques in mind, you would end up with a really long list of practices that pop. You can do this with any great content creators; they all use these techniques meticulously and generously with good reason: they work. Ask yourself what elements you already use and if you can use them better. Then take note of the elements you don't use that you need to implement in future content creation.

Content for Sales

Some of you might be thinking, "But I'm not using YouTube to grow an audience; I am using YouTube to sell." Whether you're making a video to build an audience or making an ad to sell a product, the formula is the same. Although, when you want your viewers to buy something, you make content with a problem-solution formula in mind. You must explain a problem your viewers can relate to, and when you offer a solution, you back it with credibility. The original direct marketing master was the infomercial guy, Ron Popeil, who coined the phrase, "But wait, there's more!" that people quote to this day. Ron followed the formula for making good content by putting it together in a story, even when his end goal was to get people to pick up the phone and buy a product. He engaged with the audience with a quick hook and a problem that a lot of people could relate to.

In the early 1950s, Ron filmed the first-ever prime-time infomercial for his food chopper called the Chop-O-Matic. He began with a visual hook, demonstrating the ease and speed of the "greatest kitchen appliance ever made" (there's the verbal hook, too). He proceeds by telling the folks that the Chop-O-Matic will make their cake baking, candy making, and ice-cream sundae topping exponentially easier. These specific food-making tasks were not randomly pulled out of a hat. Ron had asked crowds at product demonstrations which kitchen tasks were most problematic, and the majority of answers came down to chopping nuts for those three reasons. So he addressed the biggest problem first. He continues to use the problem-solution strategy as the video goes on, showing a remarkable speed and ease in chopping anything from celery to ice. He claims that the chopper's crowning feature is that it chops onions so quickly that it saves your hands and your eyes. No more tears!

Ron appealed to both women and men at different points in the ad, hitting his number one customer first (the wives and mothers), and his backup demographic second (the men). The visual sale worked on its own, but paired with a flawless script, discounted price, and bonus recipe book, the Chop-O-Matic was a slam dunk campaign. It made millions and launched Ron's extremely successful career. His content, like all great content, grabbed attention, explained simply, built to a climax, and had a payoff at the end. To make the sale, he presented a problem, offered a solution, and demonstrated his credibility again and again.

It Works Across the Board

You might be wondering if all of these content elements really can be adapted to any scenario. After all, a *Seinfeld* episode runs for 20-some minutes, and MrBeast's video was almost 17 minutes. How could you put all of these elements into a shorter, more focused presentation? Let's return to the Wendy's commercial, which is a mere 30 seconds long, but still has all of the elements of good content:

- **Hook**: Three cute little old ladies are at a fast-food counter. There is a large hamburger on the counter. Behind them, dull restaurant tables look very much like dull office cubicles. The contrast is so striking that it immediately pulls you in.

- **Reengagement**: The women look at the hamburger on the counter and start commenting about how big and fluffy the bun is.

- **Setup**: Ten seconds into the commercial, one lady takes the top bun off, revealing a tiny burger patty inside. Another lady says, "Where's the beef?!"

- **Climax**: Fifteen seconds into the commercial, the picture changes to show a close-up shot of a thick hamburger, and a voice-over tells you that Wendy's "modestly" calls the burger a single, and it has more beef than Burger King's Whopper or MacDonald's Big Mac.

- **Goosh**: At the 25-second mark, one woman holds up the giant bun to her ear, listening to it as if it is an empty sea shell and she's trying to hear an ocean that isn't there. At the same time, another lady says about the apparently missing restaurant staff, "I don't think there's anybody back there." That's a powerful combination of good comedy while reinforcing that most hamburger places don't care if you want a better burger.

- **Wrap-up**: In the closing seconds, the announcer says, "You want something better. You're Wendy's kind of people." The company wants to help you and the cute little old ladies, all while making you chuckle. Doesn't that make you feel good?

Simple storytelling that can be explained in one sentence: Wendy's gives you more beef.

Make a Lasting Impact

There is a pattern behind making great content, whether it's a 30-second commercial from the 1980s, a 17-minute car lot giveaway, or an episode from a "show about nothing." The greatest content creators are master storytellers who know how to make a lasting impact on an audience. They do this by following patterns and implementing tactics to hook, reengage, play on emotions, deliver on promises, have unexpected surprises and payoffs, and offer bonus content before they leave you. If they're really great, they'll leave you wanting more, and wanting to share. These core principles are universally applicable, regardless of length, target audience, and purpose.

Action Exercise

Task 1: Go back to the list you made in Chapter 12's Action Exercise. Analyze the videos using Figure 13.1 to see if they use a story arc. Identify the hook, reengagement, setup, climax, and the goosh.

Task 2: Create your next video using the story arc.

Get the companion workbook and find more resources at www.ytformulabook.com.

14 Feedback Is Queen

Content might be king, but feedback is queen. Your kingdom might muddle through with just one or the other, but you need both on the throne for your empire to thrive. You know how to make good content. Now you need to learn how to unlock the power of feedback to really grow.

We need to define feedback as it pertains to your YouTube channel. Most people probably think of human input when they hear the word feedback. You get positive and negative feedback from your partner, your boss, your mom, your friend, and your friend's mom. It is true that you should get human feedback about your content, but I'm also talking about data feedback from YouTube. Let's talk about both.

Human Feedback

First, human feedback. When it comes to your YouTube channel, you actually shouldn't seek feedback from your mom or your friend. If your mom is anything like mine, she'll tell you your video is great even if it's the worst video ever uploaded to the Internet. Or maybe you have the kind of mom who criticizes everything you do, and she would pick apart your video even if it's the best thing that ever happened to the Internet. Either way, she's the wrong person to ask. Why? Because she's not your audience.

This is why you don't start a new YouTube channel and then hound every Facebook friend and Instagram follower to go watch your new video. It's not the way to get the right subscribers and viewers. Their viewing behaviors are not the same as your ideal avatar. And their verbal feedback will sound something like, "Awesome, bro!" or "It looks great!" or "OMG, your hair and makeup were, like, totally perfect!"

A lot of creators think joining a Facebook group will be a good place to get good feedback. You have to be careful with Facebook groups, because they often sound just like your mom; people are only full of encouragement, or they're full of advice or criticism when they don't even know what they're talking about. Instead, join or create a mastermind group of trusted, relevant people in your space. You might chat weekly via a remote online meeting, or you might meet in person if you live in a place with similar creators. I have a weekly mastermind group with a handful of people whom I trust to bounce ideas off of and get good feedback from. We talk about new projects, the algorithm, problems we run into, collaborations with each other, and all things YouTube.

My group coaching program includes private feedback groups for all participants. I use Discord, but there are lots of options for online meeting spaces, like Mumble, Zoom, TeamSpeak, MeetUp, or even Google Hangouts. Do some leg work and join or start a group that suits you. If your best option is to start your own group, consider creators in your niche who you trust or aspire to be like. Or weed out the good ones from a Facebook group and ask them. It never hurts to ask. I love my private Discord group because they share commitment and passion, and it's a safe place to be open to discuss and accept feedback. They love it, too. Some great ideas have come from conversations they had in the feedback group.

I mentored a YouTuber named Steve Yeager (his channel is called *Shot of The Yeagers*). One of the first things I told him was to find two or three other YouTubers that he could meet with weekly to talk strategy and bounce ideas off of. Steve lives in Utah where there is a huge YouTube community, thanks to Devin Graham, aka "devinsupertramp," and other OG YouTubers. Devin and others started a group called "UTubers" (Utah YouTubers). I told Steve to start attending the monthly meetings and look for some creators to mastermind with. He said, "Derral, I only have 100 subscribers; no one will want to mastermind with me." I told him I knew of three other channels that would be an ideal fit for him (one had 1,000 subs, another had 5,000 subs, and one had 15,000 subs). He approached them, despite his reservations, and the two smaller channels were excited about starting a mastermind group. The channel that had 15,000 subscribers passed on the opportunity, thinking they were too big of a channel to meet with newbies.

Steve and the other two channels started to meet weekly. They strategized ideas, talked about videos, and more importantly, shared data. They also started to collaborate on videos. There was true synergy. All three of these creators' channels exploded. Today, Steve's *Shot of The Yeagers* channel has 4.9 million subscribers, and he owns four other channels. *The Ohana Adventure* has 3.3 million subscribers and four other channels. *The Tannerites* channel has 2 million subscribers and four other channels. But remember the creator who turned them down? They have 45,000 subscribers. There is true power getting in-person feedback and encouragement. I would strongly recommend that you find like-minded creators in your area and meet regularly. Start a group like Steve did.

MrBeast released a video called "Anything You Can Fit in the Circle I'll Pay For." The video wasn't getting as many clicks as a MrBeast video normally would. So he changed the video's thumbnail to include fewer items in the circle, and the video performed better but still not like he wanted it to. He reached out to our mastermind group to get human feedback, and someone suggested that he leave the circle empty.

So he went back again and took all the items off the thumbnail to show the circle empty. Viewers responded really well to this thumbnail, and the video took off. When he uploads a video, MrBeast is a pro at watching his analytics in real time and adjusting according to the data feedback from YouTube or the human feedback from his colleagues. He course-corrected with the feedback YouTube was giving him in real time, which gave him a 6% increase on CTR.

Data Feedback

Besides getting feedback from the right humans, you also must listen to the data feedback from YouTube's analytics. Feedback sets in motion the "analyze and adjust" part of the Formula. I can't stress this enough: if you don't study your analytics and course-correct based on the data you see, you'll never succeed on YouTube. The term "course-correct" is important because it's something you do while in transit. It means there is an immediate problem that needs tending to, because if you put off fixing it, you'll end up nowhere near your desired destination.

Not quite sure how to course-correct in real time with your analytics? Let me illustrate with a story. I travel all over the world to speak at events or work in person with my clients. Because I travel so much, I prefer to fly out of my local airport. However, I had a speaking gig in London once and I wanted a direct flight, which meant I needed to drive the extra two hours to the international airport in Las Vegas, Nevada. I took my son Thatcher with me so we could spend his birthday together in London and Paris. Our

10-hour return flight landed us in Vegas late at night, and I was jet lagged and ready to be home. So I zipped up the interstate as fast as I dared, when along came a construction zone and its inherent reduced speed limit.

Now, I already told you about the long flight and the jet lag and the homesickness, right? Well, those things added up to me not wanting to slow down. The road was smooth and straight, so even with several signs warning me that fines double in a work zone and that the speed limit was now 45, I kept cruising. It wasn't long before I saw those dreaded red and blue flashing lights behind me. BUSTED! I had earned myself a hefty ticket and a seat in defensive driving school. We would have been better off with Thatcher driving, and he wasn't even old enough to have a license at that time.

A month later, I was back on the freeway in the same scenario coming back from Europe: in a hurry to get home with a smooth, open road. This time, however, there was something in addition to the same old construction zone signs. It was a flashing sign that said, "speed limit 45 . . . your speed 65." This sign gave me real-time feedback to course-correct. Yes, I could have looked down at my odometer to see that I was going 65 miles per hour, but I didn't. I couldn't ignore the flashing sign, because I knew the consequence of speeding in a construction zone. So I slowed down, and didn't get pulled over.

YouTube gives real-time feedback just like that flashing speed limit sign. Every time you upload a new video you get feedback in your analytics to help you immediately course-correct what's not working great. The AI is very sensitive to your traffic sources. It tracks impressions and clicks. The click is key to your adjustment strategy. Click-through rate (CTR) lets you see what's happening on your video in real time so you can quickly pivot as needed.

The Four Ws

YouTube has given creators amazing and invaluable tools in YouTube Studio. These tools help us really understand the viewer inside and out and also quantify our videos' successes and failures. YouTube analytics helps you break down the elements in metrics so you can course-correct. The YouTube development team has done a great job here. It seems every week there is a new metric or tool added to simplify their reports. I am thrilled with the robust analytical tools and insights. Numbers and graphs are intimidating for some people, so they just stay away from their YouTube analytics altogether. Don't let "analysis paralysis" keep you from learning. Getting to know your data is a crucial part of the YouTube Formula.

To begin, take a step back and don't overcomplicate what we are trying to analyze. We need the feedback that will help use reach and connect with the audience. We need to consider the basic information gathering questions, which we'll ask with the Four Ws: Who, Where, What, and When.

We start with *Who* (the Audience). This is a lot more than just the simple demographic data. We can break down your audience in your Channel analytics to show your unique viewers, how many videos they watch, your subscriber growth or decline, when your viewers are on YouTube, and how many of those viewers are subscribers who have bell notifications turned on. You can see your viewers by age, gender, and the top countries and languages represented among them. See Figures 14.1 and 14.2.

The second W is the *Where*. Your traffic source report is found under the Reach tab. It will help you uncover where the viewers are watching your video: from within YouTube, or from external sources or embeds on websites. See Figure 14.3. You can see which traffic source is bringing the most views, visibility, and Watch time. Understanding which content YouTube loves to recommend (Browse

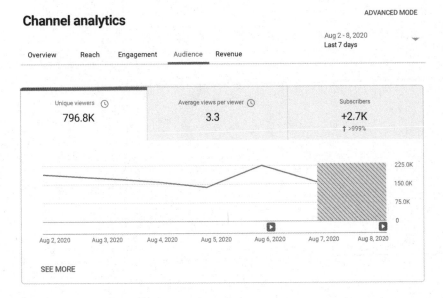

Figure 14.1 The Who: unique viewers

Figure 14.2 The Who: viewer demographics

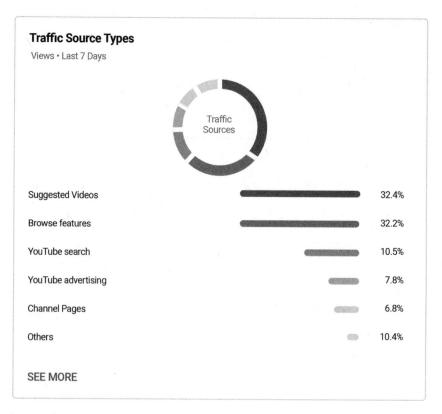

Figure 14.3 The Where

features, Suggested videos) will help you understand your content strategy better and help you boost viewership. This is by far one of the most overlooked metrics on YouTube, but it is crucial for your growth strategy.

The third W is the *What*. The What is your video the viewers are seeing and clicking on to watch. It includes impressions, Watch time, CTR, average view duration (AVD), and average view percentage (AVP). See Figure 14.4.

Last is the *When*. I break the When into three parts: real time, date range, and when your viewers are on YouTube. The real-time

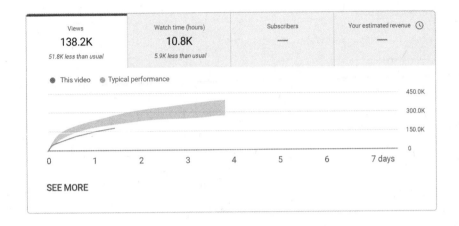

YouTube recommending your content

Compared to your other videos, a smaller group of people are watching this video from recommendations

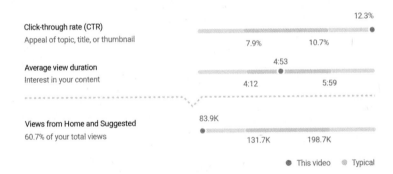

Figure 14.4 The What

metric shows how your videos are performing in the last 48 hours, as well as the last 60 minutes. These are live analytics. Figure 14.5 shows the date range, and Figure 14.6 shows when viewers are on YouTube.

With the Four Ws, you can start to see how viewers respond to your content. It will give you the data you need to course-correct without making things overly complicated.

Figure 14.5 The When: date range

Human + Data Feedback: A Winning Combo

A coaching student of mine, John Malecki, has a wood- and metal-working channel. His thumbnails used to feature just an object—the beautiful finished product of his work. John was very hesitant about putting himself in the thumbnail, but I convinced him to give it a try because data shows that thumbnails perform better with people's faces in them.

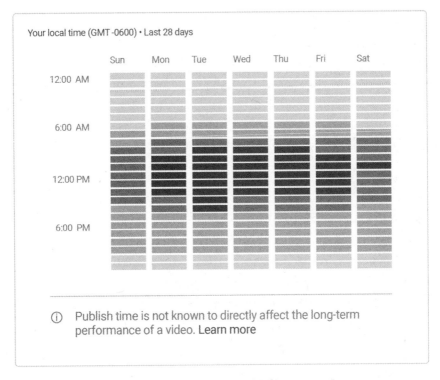

Figure 14.6 When viewers are on YouTube

As an ex–NFL lineman, John thought he was too big to be in the picture. Then he decided to go for it anyway; he started making more personality-based videos and putting himself in the thumbnails. Viewers started clicking more and watching longer. John had made a video about a DIY lava table and the thumbnail featured only the table. Three days after uploading it to his channel, his CTR was low at 1.6%. So he switched out the thumbnail for one that had him in it with the table. Some of his previous thumbnails showed John pouring something and had gotten a lot of views, so we thought he should use that tactic here. He poured the "lava" into a bucket, and an arrow pointed at the finished product. Then we watched his CTR—it immediately jumped up to 8.9%. A video that was on its

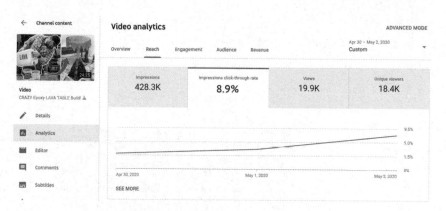

Figure 14.7 John Malecki CTR

way to being one of his lower performing videos is now a top performer . . . all because John sought out human feedback and data feedback and made adjustments from both. See Figure 14.7.

Don't worry if you've never watched your analytics in real time. Don't worry if you've never made adjustments to your content based on the feedback you're getting from the data. The good news is you can start doing it right now.

I'll show you exactly how in the next several chapters. We'll cover CTR and other metrics like AVD, AVP, and average views per viewer. We'll also go over your content strategy and how to categorize your videos. You'll be able to create awesome content and make smart, data-driven decisions to help that content perform its best. That's a powerful, channel-ruling combo!

Action Exercise

Task 1: Look in your area for other YouTube creators. See if there are any meet-ups already happening, either online or offline, and ask to join.

Task 2: If there aren't any, create your own mastermind group and meet regularly.

Task 3: Look at the videos you have released in the last 90 days. Write down your top-performing videos and analyze each one using the Four Ws. Look for patterns among them.

Task 4: Validate your viewer's persona as identified in Chapter 11's Action Exercise. Are they who you thought they were? How are they different?

Get the companion workbook and find more resources at www.ytformulabook.com.

15 Title and Thumbnail: Success Starts with a Click

What is the most important metric for YouTube success?

This is one of the most frequently asked questions I get year after year. The answer is easy, and it goes something like this: everything starts with a click. If the people don't click, they won't watch, no matter how amazing your video is, end of story. How do you get viewers to click? You earn their attention. Getting viewers to click is the biggest struggle I see for creators and brands alike. You have to get people to focus on your content outside of all distractions.

My son Logan decided to run for student body president at his high school, so he asked me to help him with his marketing. Of course, I was thrilled to help; I love new campaigns and talking strategy, but this time it was for my son. We talked about the importance of grabbing people's attention with a message or slogan. It needed to be easy to remember and simple to explain and share, and it needed to stand out above the other campaigns. He said, "Dad, you know I hate attention. I hate the spotlight." It's true; since he was a little kid,

Logan was always uncomfortable with attention. I pushed him a little, saying maybe he needed to get out of his comfort zone. But then he said, "I don't want to be fake. I'm just me." *Boom.* An idea struck me. His campaign messaging would be, "No Slogan, Just Logan." It effectively grabbed people's attention and was easy to remember. The slogan reminded his peers that they liked him because of who he was, and that's what won him the election in the end—not the campaign.

In earlier chapters, you learned that YouTube's job is to predict what the viewer will watch. The AI is really sensitive to everything that viewers engage with on the platform, and it actually starts with an impression. An impression happens when a viewer is able to see a title and thumbnail for at least one second. This could be on the YouTube Homepage, Subscription feed, in Search results, or Suggested feed (Up next). If the viewer skips by the title and thumbnail quicker than one second, it doesn't count as an impression. If they click, that's what records your click-through rate (CTR). An important note: the higher the impressions and CTR data the better. That means YouTube is recommending your video out to a more general audience. The more it recommends, the more impressions you will get. The more impressions you get, the lower your CTR. So don't freak out that your CTR percentage is dropping when your impressions and views go up. That's a good thing because YouTube is showing it to more viewers outside your "normal audience."

Before YouTube ever gave us CTR data in Creator Studio, I knew that it was really important just based on how I thought the algorithm worked. Success rates for ad campaigns focused so much attention on CTR data. I began split testing thumbnails and titles in AdWords for my biggest clients. We put in our audience demographics and interests and created an ad to see which one had the highest CTR. I was surprised when we changed out those winning titles and thumbnails and put them on YouTube. The videos would explode! I began begging the dev team at YouTube to give creators that data. I knew it was a huge factor, so I continued to spend a lot of time, money, and energy learning about which titles and thumbnails worked because this was the only way to get people to watch or buy.

Let me give you an example of how to do this. When Jeffrey Harmon, Daniel Harmon, and their team and I created the Squatty Potty ad, we made a Google doc for title ideas and added every variation we could think of. There are no "dumb" ideas in brainstorming, because anything can inspire a great title. We needed to figure out what was going to pique human curiosity and make them click. Because more clicks would bring us more people taking action to buy. We knew we had to test our options, so we did 165 title variations to see how viewers responded to them. We all had our favorites, but we had to choose based on the data—the data doesn't lie! And we really wanted to get it right because we had a lot riding on this campaign. Success was our only option. The best way to see which title and thumbnail brought the highest CTR was to run Facebook engagement ads. We created several ads for Facebook to figure out how our target audience responded, running A/B split testing on different titles and scenarios.

Interestingly, the very first title I had come up with eventually became the final title, "This Unicorn Changed the Way I Poop," but it took a couple hundred dollars to know that it was the best performing title. The inspiration came from looking at a list of Daniel's titles and brainstorming from there. That's the power of collaboration and brainstorming together. After selecting the titles, we needed to start testing thumbnails. We also took 20 different thumbnails and tried them with our top five title options, spent a couple hundred more dollars on additional split testing, and made our final decision based on the data. Our final CTR was 10 times higher than it would have been without testing. Ten times. Consider that for a moment.

Jimmy Donaldson, "MrBeast," has become a wizard at title and thumbnail creation for optimal click response. I love Jimmy's process of creating the best title and thumbnail for each video he makes. If he can't come up with a clickable title and thumbnail, he won't make the video, no matter how good the video idea is. Ever! Trust me, I've tried to convince him to. Most creators think only after shooting the video and right before uploading, "What should I do for the title and thumbnail?" This is a huge mistake and it could cost you the video's potential success. Put a lot of thought into what people will want to click. This is critical because if they don't click, they won't watch. That's why Jimmy strategizes before he makes every single video. We literally have spent days coming up with the right title and thumbnail. Not hours, days. Let that fact make you stop and think about your title and thumbnail creation process and priority. Make it a priority. You don't need to go to the extreme like MrBeast does, but you should be thinking about your potential viewer and what would motivate them to click. This will help you when you are creating the video as well. Viewer first!

Grabbing Attention Visually: The Science

In my college biology class, I learned about the brain and the visual cortex, and I was fascinated with it. My interest grew when I could see practical application in advertising. I spent hours trying to figure out how it applied to grabbing people's attention. To summarize, the visual cortex of the brain is what processes visual information. There are four areas to the visual cortex called V1, V2, V3, and V4. There are also special areas of the visual cortex that process visual information very quickly. These are called the "preattention" areas because they process information faster. Let me show you how this works.

Let's do a quick exercise called "the blink test." Close your eyes and then open them for a millisecond and look at Figure 15.1. In other words, just blink and then keep your eyes shut for a few seconds.

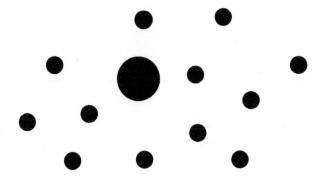

Figure 15.1 Sizes and shapes

Notice where your eye was drawn. Was it drawn to the bigger circle?

Let's try it again with Figure 15.2.

Was your eye drawn to the slash that was a different orientation?

These examples are super simple but they work. Science rocks! You can't get the full power of the process in a black and white book because it works best with color images. The visual effect of color in thumbnails is so important that I created a free online course to go along with this book. You'll find the best color thumbnail examples, along with my detailed thumbnail training, at ytformulabook.com.

Figure 15.2 Orientation

The preattention areas of the visual cortex are not the only visual/ brain connection we use. Another area of the brain you can tap to grab attention on a thumbnail is the Fusiform Facial Area (FFA). This part of the brain is sensitive to human faces and is close to the area of the brain that processes emotions. This serves as an explicit trigger in thumbnails because the brain has circuits devoted to faces. When my daughter Ellie was four months old, I was reading a study about how babies can recognize faces and emotional expressions starting at about three months old. Of course, I had to put it to the test. I tested six facial expressions on baby Ellie to see what emotion it produced. See Figure 15.3.

Fearful   **Angry**   **Sad**

Happy   **Disgusted**   **Surprised**

Figure 15.3 Expressions

For weeks I tested my "crazy" faces on my baby. I was amazed to see that each facial expression created a different emotional response from even a four-month-old. I think I scarred my daughter for life (sorry, Ellie!), but it was in the name of science. I still mess with my daughter sending crazy faces to her via text to this day. Knowing how the human brain reacts to images can give you a massive advantage when designing your thumbnails.

Netflix has some of the best thumbnails on the planet. They spent a lot of money trying to perfect the science of grabbing viewer attention. I regularly browse Netflix to look at the thumbnails that really grab my attention. I never click on it to watch, I only browse. It drives my wife Carolyn nuts. She'll say, "Are we going to watch something, or am I going to watch you look at thumbnails all night." I'm fascinated by what shows up. I also go to Carolyn's account and my kids' accounts to see what shows up differently for them. I study the thumbnails a lot. I even create test accounts just to see other types of thumbnails shown and what is recommended. In doing research for clients, I ask them to list 20 movies their viewers would watch, then I select them on Netflix and see which recommendations and thumbnails populate from Netflix's AI. This always gives me ideas of which colors and image positions really connect with a demographic.

Imagine seeing 12 images flash on a screen in a fraction of a second. Do you think you can process each image? A team of neuroscientists from MIT have found that the human brain can process entire images that the eye sees in as little as 13 milliseconds.

There is a Netflix research article called "The Power of a Picture" that reinforces the importance of an image to help a consumer choose what to watch. In their research, Netflix found that the artwork associated with a movie accounted for more than 80% of people's attention while browsing. Attention given to titles was secondary and lasted less than two seconds. That's not a lot of time to convince with words that this is what that person wants to watch. The image is most important. And it's got to be good, because we see things in the blink of an eye and make snap judgments without even realizing it. As content creators, we only have milliseconds to grab people's attention and pull them into our content. If we don't grab them fast, they've already moved on to look at something else that will get them to click. How do we grab their attention? With curiosity.

Curiosity is a basic human behavior—we are programmed to be learning machines. It's a natural human instinct that has been with us since the day we were born. It never really hit me until my daughter Ellie was three years old and was always asking questions. She asked classics like, "Why is the sky blue?" and, "Where do babies come from?" but my personal favorite was, "Mommy, why is Daddy so loud?" Curiosity is a fundamental element of our cognitive functioning.

Curiosity earns the click. Your thumbnail and title combination should make someone scrolling through pause long enough to consider clicking. For example, when I saw a photo of a guy holding a loaded barbell underwater, it grabbed my attention so I took a quick glance at the title, "Can You Bench Press 1,000 Pounds Underwater," and I was hooked. See Figure 15.4.

Figure 15.4 Bench press

I had to click on that video. I unintentionally spent the next hour and a half consuming creator Tyler Oliveira's content. Tyler made clickable thumbnails with engaging titles that piqued my curiosity over and over again. Plus, his video content did not disappoint. Let's cover some basic design principles that will enhance your CTR percentage.

The Rule of Thirds

To make sure your image draws in the human eye, keep it simple and balanced. A viewer feels more connected to a well-balanced image because it feels aesthetically right. An image that appears static will always look less interesting. One of the first techniques a beginner photographer learns in a photography class is the "Rule of Thirds." It's an effective technique to compose a balanced photo by dividing it into thirds, both horizontally and vertically. The subject of the image is positioned at the intersection of those dividing lines or along one of the lines itself. The dividing lines of the Rule of Thirds effectively breaks down an image into nine parts, as shown in Figure 15.5.

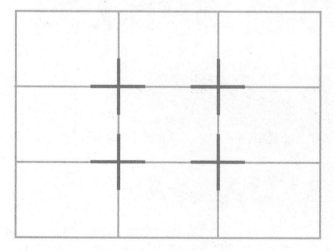

Figure 15.5 The Rule of Thirds

Studies have shown that when people view an image, their eyes naturally go to one of the intersection points. It's easier for the brain to process quickly. It's good to learn about natural processes like these so you can learn to work with them rather than against. Figure 15.6 shows three examples of the Rule of Thirds. Once again, to get the full effect, close your eyes and then open them when you look at Figure 15.6. Take note of what your eye naturally focuses on first.

Always ask these questions when taking photos or designing thumbnails:

- What are the points of focus in this shot?
- Where do you place that focus in the grid?

With all of this said about "The Rule," don't let it stress you out every time you go to make a thumbnail. As they say, rules are

Landscape

Face

Object

Figure 15.6 Three examples of the Rule of Thirds

meant to be broken, not every time, but sometimes you'll get a better shot without the grid anyway. You can always edit later with post-production editing tools if you need to, cropping and reframing to get your image to fit within the grid.

Color Psychology

I watched a YouTube video about a 10-year-old boy named Cayson Irlbeck who was born color blind. Color blindness is widely misunderstood. Many people assume it means that someone can only see in black and white, but it doesn't. It means the brain can't process different shades of color. In the video about Cayson, his dad wanted to surprise him with something, so he had him go outside on a beautiful day. The sky was bright blue and the grass was a luscious green, but Cayson didn't see it that way. His color-blind brain saw a blue or purple sky and red grass. Cayson's dad then gave him EnChroma glasses. These glasses allow color-blind people to see colors correctly, something Cayson had never experienced before. He put on the glasses, and tears started running down his face. His dad cried, too, both sobbing for joy. I wish everyone could see Cayson's face. I'm so grateful to live in a world with color.

In 1665, Sir Isaac Newton was studying about light and color at Cambridge University. He was observing how light was reflected into various colors. On a sunny day, Newton darkened his room by covering up his window with drapes, making a small hole to allow just one beam of sunlight into the room. Then he held a large glass prism in the beam of light. He was able to see the full color spectrum. This led to more prism experiments, not only by him, but by countless scientists studying color and light.

Color plays a huge role in influencing decisions. We call this color psychology, and it can be a powerful tool of persuasion. I literally wear a black T-shirt every day of my life, but don't let that fool you for a second into thinking that color doesn't matter to me. When it comes to YouTube content, especially thumbnail images, I know that color can make or break your video's success. Poor color choice in a thumbnail will negatively impact your video's clickability. A study published in the journal *Management Decision* by Satyendra Singh says that color alone contributes up to 90% of the information that forms a decision. Ninety percent? If this is true, it's extremely important to keep color in mind when creating thumbnails. The color is essential to earning that click.

Everyone learns about the color wheel in grade school. When I learned it, I didn't know how much it was going to help me in my career. Contrasting or complementary colors are opposite each other on the color wheel and can be used together as powerful combinations to make your images stand out. Keep in mind that you can still be contrasting with grayscale but it would feel bland just like Cayson's color blindness. I will show you the example in grayscale in Figure 15.7, but remember, you can see it in color in the free online course.

High- vs. Low- Contrast

Figure 15.7 Contrast in black and white

Color is an essential tool in your thumbnail image creation process because it has a huge impact on how viewers think and behave. This process has been studied and analyzed by marketers and advertisers, because it helps predict human behavior. Understanding it means you can make content more impactful. The visual cortex V1 that processes color directs the viewer's eye where to look, what to do, and how to interpret something. It also can add context to content. It helps us decide what's important and what's not. That's precisely why, as a content marketer, you need to understand what colors mean to people.

Thumbnail Strategies

We've covered some of the science behind why thumbnails matter so much: how creators capture viewers' attention with photography techniques and the right color combinations. Now let's dive into strategy. How do you use this information to your advantage?

Before 2012, YouTube would auto-select a random clip of a creator's video and set it as the thumbnail image. If the video's creator didn't like the auto-image, they had to re-upload until they liked the option YouTube chose. I know, because I used to do this. I knew how much that thumbnail image affected my CTR, so I did what it took to get it right, tediously re-uploading until I liked the thumbnail.

YouTube rolled out its Partner Program (YPP) in 2012 and quietly added new features to help their YPP creators. Creators who belonged to the Partner Program had a huge advantage with these features, one of which was being able to upload a custom thumbnail image. YouTube sent out an email to their YPP members encouraging them to use this new custom thumbnail feature. These select YPP members—there weren't a lot at the time—now had a huge advantage over the general YouTube public. It didn't take long for many of them to realize that they could make sensational thumbnail images to get lots of clicks and views.

This is when the "clickbaiting" tactic hit YouTube the hardest. YPP creators used baiting techniques like bright colors, arrows, emotionally charged facial expressions, and big cleavage shots to get viewers to click on their videos. They were being deceptive and misleading. Viewers took the bait and clicked, but it didn't take long for them to realize that the content didn't show what was advertised in the thumbnail. YouTube noticed trends in the data from the new custom thumbnail feature, so they started using different metrics to inform their Suggested video recommendations. This helped to minimize the clickbait problem (see Chapter 2 for more about this issue). Note: Clickbait can be good if your video content delivers on the promise and isn't misleading the viewer.

I watched custom thumbnails transform YouTube. As a marketer, I observed some things with thumbnails that I had never seen before in advertising. People got really good at learning thumbnail design strategies to get people to click more. These are simple strategies that any creator or brand can use, and I'll show you how to use them in Figure 15.8.

Data has shown that the best thumbnails include an object and a person. When you drill down on that Netflix research finding that we mentioned earlier, it shows that viewers engage with thumbnails that have people in them. The thumbnail should be able to tell a story without words. You can do this effectively with close-ups and complex emotions on the face of the person or people in the image. Figure 15.8 shows some of the most common types of thumbnails on YouTube. Keep in mind that these examples are in black and white, and color really makes these things pop and make sense. I encourage you to go through the detailed thumbnail training at ytformulabook.com.

Face First In a face-first thumbnail, your facial expression is the most important thing. The best way to connect with someone is through the eyes. So make sure your face is up close and your eyes are nice and bright.	Figure 15.8a 
Face and Object If you have an object to show, this strategy is a great way to pull in your audience. The emotional reaction on your face along with an interesting object will really make your thumbnail stand out.	Figure 15.8b 
Object First Sometimes the object is the star of the show. In this case, your object should be right in the center. Really big, really bold, and really bright. Make sure there is no confusion to your audience on what to look at.	Figure 15.8c 
Two Faces If you plan on having two people in your thumbnail, you should have the faces on the left and right of the frame with the object in the middle. Don't forget to show emotion!	Figure 15.8d 

Three-Panel A three-panel thumbnail is perfect for showing progression. Your eye goes from left to right, so be sure to put the "before" on the left and "after" on the right. Remember that progression could mean time, but it could also mean change. Different colors can also help enhance the effectiveness of this thumbnail.	Figure 15.8e 
Two-Panel The two-panel thumbnail is another way to show progression. The perfect example is "before and after." You can also frame it as "right or wrong," or "real or fake." The juxtaposition is what is important in this type of thumbnail. This type not only can apply to objects, it can apply to people's faces as well.	Figure 15.8f 
Text to Amplify Text can be super powerful when used sparingly. Text should only be used to help the audience's understanding or raise their curiosity. Keep text simple. If you can't tell a story with just your thumbnail, use text to clarify.	Figure 15.8g 
Clipart Use red arrows, circles, hand gestures, and emojis to redirect the viewer's eyes to a specific person or object in the thumbnail.	Figure 15.8h 

Perspective

Our minds are drawn to extremes, especially when it comes to really big things or really small things. Showing scale in your thumbnail in relation to you can help make your thumbnail really pop.

Figure 15.8i

Organized Clutter

If you have a lot of one thing you can use the "organized clutter" strategy. Placing yourself in a uniformly cluttered background can help you pop. Make sure there is a pattern in your background and not just clutter—the last thing you want is to have a messy thumbnail.

Figure 15.8j

Showing Action

These thumbnails create the "I wonder what's going to happen next?" effect. Curiosity earns the click, and this thumbnail strategy should always create curiosity. People love action movies and action shots, so when you can, show action.

Figure 15.8k

Colors Draw in the Eye

Each color has an emotion associated with it, so how you use it can influence how your audience feels about your thumbnail. Using the right color combinations can also help draw the viewer in. Some strategies are to use complementary colors or colors on the opposite side of the color wheel. That contrast will really make your thumbnail pop.

Figure 15.8l

Figure 15.8 Thumbnail types

When you're designing your thumbnail, always keep in mind that the majority of YouTube viewers are watching on mobile devices now. You want to ensure your thumbnail looks clear when viewed on mobile. A lot of creators design on a computer, but you need to view on mobile to see what most viewers see. Ask these questions: Are there any objects that are too small? How about things that you can't easily distinguish? And do the blink test, closing your eyes and opening them to see what grabs your attention first. Super Tip: I always want to see how my thumbnail looks compared to other thumbnails. To do a mock comparison, screenshot search results and photoshop your thumbnail in with the list of results. This can be super powerful to see how the viewer would see your thumbnails.

Thumbnails the Right Way

To do a thumbnail the right way, plan ideas before you even shoot the video. Ask yourself, "How do I explain what's going on without words?" Then go through the thumbnail brainstorm list and come up with ways you could use two or three different methods to design images for this particular video. Often it takes a combination of these tactics to capture viewers. Remember that the more emotion you put in the thumbnail, the more intriguing and clickable it will be. Then do a photo session before shooting the video. Take pictures from different angles, and remember that simplicity sells.

One extra word of advice here: don't take a screenshot of someone else's thumbnail with the intent of copying it on yours. It's just wrong! The more you brainstorm consistently, the easier it will be to get your own creative juices flowing.

Thumbnail Audit

You should have two or three variations of each thumbnail concept you choose, because you need options to test or to quickly change out if you are not getting a good CTR. Once you've completed the photo shoot and the images are edited and ready to go, you can go through your thumbnail options and ask yourself:

- Which image is more clickable?
- Does it actually portray the video content?
- Does it get the audience excited or intrigued?
- Would *you* actually click on that thumbnail?

Traffic Sources and Thumbnails

I've been known to create several types of thumbnails for the different stages of viewer consumption. I realize this is super nerdy and tactical, but I'm really sensitive to where the impressions are accruing (Traffic Sources) and what the CTR baselines are. On some of the biggest videos I've worked on, we designed thumbnails that would get more clicks on the Browse feature (YouTube Homepage and Subscription feed), then we would switch out that thumbnail to a different thumbnail designed for higher CTR for Suggested videos (Up next). Yes, there is a difference even on that granular level. I've seen a triple increase in views by using this strategy. It can help you if you will look closely at those CTR numbers.

Video Titles

Netflix's study told us that once we grab the viewer's attention with the thumbnail, we have 1.8 seconds to make an impression with our titles, so let's not waste it. This can be the difference if someone will be watching your video or moving on to the next thing that grabs their attention. So, what makes a good title?

In Chapter 11, we talked about how to identify your audience. We talked about learning their offline and online behaviors. This is super important when it comes to titles, because you don't want to use words or phrases that disconnect and confuse your audience. I remember one time my son Kelton was talking to his older brother Logan about skimboarding with some friends. Kelton told Logan about a "sick trick" he landed and how "dope" it was. My mom, a baby boomer, was in the room, and she turned to me and said, "Is Kelton okay? Is he sick? Are his friends doing drugs?" After laughing uncontrollably for a few minutes, I had to reassure her that Kelton was healthy and his friends weren't doing drugs. The words he had used were disconnected from his baby boomer audience. Certain words, like "sick," and "dope," meant something entirely different to her.

You have to consider the subconscious split decision a viewer is making when they look at your thumbnail and then title. In that split second, their brain is asking:

- Is this video what I'm looking for?
- Do I really want to watch this video?
- Is this video going to be worth my time?

Whether you like it or not, your YouTube titles help shape the viewer's decision to click.

Your Title Needs to Satisfy the *Why* and the *What*

Your goal for your video's title is to reinforce the thumbnail. It gives added context to what initially grabbed the viewer's attention. Make all titles clickable for humans and try to predict how your audience will respond. The title should be easy to remember, simple to explain, and easy to share. I want to take you through an exercise that I call Macro and Micro. This exercise has been super helpful in solving problems in my career, but more importantly, in helping understand people's intentions and behavior.

Let's look at the macro level first, the *Why*. Why is the viewer coming to YouTube? Why are they scrolling on the Homepage? Why are they searching videos on YouTube? Why are they watching these videos or not watching those videos? There was a 2017 study called, "The Values of YouTube," that tried to validate some assumptions about why people come to the platform. I always geek out over any new study Google does relating to YouTube, but geek or not, every YouTuber needs to look at the study shown in Figure 15.9.

The YouTube viewer's Why is simple, and it comes down to four things: Entertainment, Education, Inspiration/Motivation, and De-stress/Relaxation. This is simply human nature.

The micro level is the *What*. What do they need to learn? What is the solution to their problem? What would they find entertaining? What will inspire them? What will help them de-stress or relax? Most importantly, what would encourage them to click? What is your value proposition to get people to click and watch?

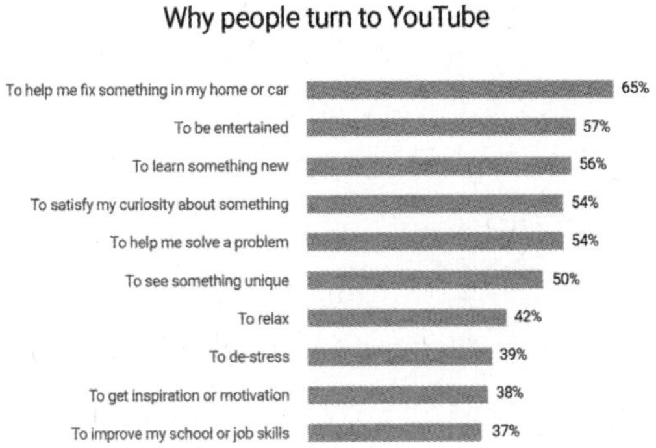

Why people turn to YouTube

To help me fix something in my home or car	65%
To be entertained	57%
To learn something new	56%
To satisfy my curiosity about something	54%
To help me solve a problem	54%
To see something unique	50%
To relax	42%
To de-stress	39%
To get inspiration or motivation	38%
To improve my school or job skills	37%

Think with Google

2and2/ Google, "The Values of YouTube" Study, (n of 1,006 consumers between the ages of 18054, with 918 monthly YouTube users), respondents were asked to choose which platforms they turn to for a range of needs, Oct 2017.

Figure 15.9 The value of YouTube

Remember this is about people coming to YouTube to satisfy their Why. The What is the keyword or phrase that will be used in the title to get them to click. This is also the topic of the video. The What is the final trigger that clicks in their brain to actually take action. The right keyword or phrase helps with the decision making process and makes the video more enticing to click on. This is about people, not the algorithm.

Thumbnail and title is where I spend the most time in my content strategy because I need to know what will get the viewer to click. I differ in philosophy with most YouTube educators in that Search volume is not my main focus but only one of many factors I use to make a decision. I would rather not spend hours doing keyword research and use those hours to focus on the viewer's What and Why, learning how people respond to my content and developing a YouTube Sixth Sense like we discussed in Chapter 10. I spend that time putting my titles and thumbnails to the test, keeping them easy to remember, simple to explain, and easy to share (I'm emphasizing this concept for the third time because it's that important). I want to spend my time thinking about how this video or project can produce a water cooler moment (a moment that people will use to discuss the video with their family, friends, and coworkers).

I try to imagine how the viewer will talk about the video after watching it, even creating imaginary conversations in my mind. . .. "Did you see that MrBeast video where he ate the world's largest slice of pizza?" If the title is easy to remember, simple to explain, and easy for the viewer to share, you've given your video a better chance of success. This is powerful. Make sure you take the time to imagine how the viewer would talk about your video and see if it passes the test or not. If it doesn't, come up with another title. If it does, then upload it and validate if people responded by looking at the data, seeing if the CTR increased.

The Macro and Micro exercise will help you dive in and get to know your viewer personas and content as we discussed in

Chapter 11. Understanding what viewers want or what they care about gives you an advantage when they are making the quick decision to click.

Making Clickable Titles

To make a clickable title you must satisfy the Why by providing value in the What. This is something that most creators struggle with for years. It can be a major stumbling block to getting people to connect with your content. I always have a brainstorming session on title ideas like what we did with the Squatty Potty ad. Then I refine the title to make it more compelling and clickable.

Below are some tips and examples on how to make titles more compelling and clickable. Let's pretend I just created a new YouTube channel called "Derral 42." My content will be about space, exploration, and the wonders of the universe, so my title examples follow accordingly. (By the way, 42 is not my age! If you are my viewer persona, you understand my reference to 42. So, don't panic.)

Use active voice. Active voice titles are effective on YouTube because they are clear, to the point, and can be easily understood. You will see that examples in these tips use an active voice. Having an active voice title also creates a sense of urgency naturally. I'll talk about urgency later in this list.

Be relevant, trending, and topical. Using trends and current events helps titles become candy for the right viewer. For example, in the summer of 2020, SpaceX was in the news for sending NASA astronauts to the international space station for the first time in nearly a decade. So if I created a video around this trending topic, my titles might be:

- SpaceX Helped NASA Astronauts Become Cool Again
- Elon Musk Helped NASA Astronauts Become Cool Again

Use questions (both open- and closed-ended). People use You-Tube as a search engine and they will ask specific questions. You can look at common questions viewers have about your topics and see what auto-populates in YouTube and Google search results. I always try to look for something—a fact, place, or person—to amplify the question and pique curiosity. Video titles for my channel might be:

- How Would YOU Survive on Mars?
- Could You SURVIVE on Mars? Elon Musk's Genius Plan
- Why Are Mars Sunsets Blue?

Include numbers for listicle video titles. This is by far one of the most powerful title and video strategies because it gives the viewer a clear idea of what to expect, a clear time limit, and helps viewers remember. Listicle titles for my new channel might be:

- 7 Essential Reasons Why Humans NEED to Colonize Mars
- 5 Extreme FACTS about Black Holes

State a problem and offer a solution. Problems always bring in human emotions. This can amplify the desire to click if you can offer a solid solution. The title would promise to show what the solution might be. Possible problem/solution titles for videos on my channel could be:

- Space Junkyard . . . Here's How We Clean It Up!
- Elon's Satellite Internet Is a Mess. This Is How You Fix It!

Create urgency. As humans, we have the fear of missing out, or "FOMO." You want to be able to offer information that the viewer will want to see right now. Creating a sense of urgency triggers the human fear of being on the outside. An urgent title for my channel could be:

- We Need to Colonize Mars NOW Before It's Too Late!

Use a trusted source. This can help those viewers that need validation to push them over the edge for the click. For example:

- NASA Says Pluto Is NOT a Planet

Address the viewer. When you address the viewer in the video title, it makes them more willing to click, because it feels like the video was made for them. Also, it adds another implied emotional context to the title. For my channel, I could use:

- Chinese and Russians Will Beat SpaceX to Mars
- Space Tourism: You Can Book Right NOW!

Use emotional drama or polarizing words: There is a reason why people become curious onlookers at car crashes, fires, and fights. Most people are naturally drawn to drama, gossip, fights, injury, controversy, sexual innuendo, illicit behavior, and opposing forces. Bringing out the inherent drama in your video title can trigger an emotional response. Drama titles for my channel could be:

- Elon Musk Calls Jeff Bezos a Copycat
- Not Made on This Earth, Pentagon Reveals UFO Findings

Capitalize your titles or words. All titles should capitalize the first letter of each word, and sometimes you should use all caps. Using all caps is one of the easiest ways to grab attention in your title. A word of caution here: sometimes all caps just feels like someone is yelling. Be calculated in your all caps usage so that you emphasize keywords without it feeling like you're just yelling all the time. Notice how I used all caps in the previous title examples, as well as the following:

- LIFE ON MARS? Neil deGrasse Tyson Thinks So
- THE EARTH IS NOT FLAT

Amplify with attention-grabbing words. Amplify the keyword or phrase to take your title over the edge. Use words like: Ultimate, Worst, Best, Faster, Insane, Crazy, WOW, or I Cried. Good amplifying words for educational channels include: DIY, Easy, Step By Step, Simple, Amazing, Quick, and Now.

Downloadable Tools for Title Generation

Over my career, I've found several tools that really speed up the process of keyword and phrase generation for titles. These tools have saved me thousands of hours of research and have helped me with brainstorming. I have a special section in the complementary course to show you my keyword and title process. Get access to the course at www.ytformulabook.com. Don't go down the SEO rabbit hole; use these tools to give you data and ideas to focus on the Why and the What.

TubeBuddy shows you what search term is ranking, the search volume (searches per month), competition for the key term, and related searches to help you dig deeper. TubeBuddy has saved me so much time! It's my favorite channel management tool for the research and optimization phase. I also love the tools TubeBuddy has for testing. Download TubeBuddy for free at www.Tubebuddy.com/go.

VidIQ Boost is designed to help you find topics and keywords for your videos. It has the most robust YouTube SEO features and competitive analysis tools I've used. They also have a feature to help you optimize your videos and give them a boost. Download VidIQ for free at www.vidiq.com/go/.

Kickass Headline Generator is one of my favorite tools to brainstorm. You can put in topics, desired outcomes, undesired outcomes, audience, points in content, and helpful aides, and it will auto-generate headlines for your video based on your parameters. Download this tool at www.sumo.com/kickass-headline-generator/.

Portent Content Idea Generator isn't as robust as Kickass Headline Generator, but you can put in a subject or topic and it will come up with clickable title ideas. I use this more for brainstorming and thinking differently. Download it at www.portent.com/tools/title-maker.

Action Exercise

Task 1: Using what you learned in this chapter, brainstorm 10 new title ideas for your next video. Narrow it down to three possible titles. Make sure your titles are easy to remember and simple to explain and share.

Task 2: Brainstorm three or four possible thumbnail ideas for each title. Use YouTube and Google images to come up with ideas to make the ideas stronger. Pick a thumbnail strategy and sketch it out (don't worry about being an artist).

Task 3: Get human feedback on these ideas from your mastermind group.

Get the companion workbook and find more resources at www.ytformulabook.com.

16 Engaging Viewers' Attention So They Watch More

You have created an awesome title and thumbnail and gotten the viewer to click, made good content, gotten feedback, and you know where your viewers are coming from . . . so, what's next? Next, you focus on one thing: retention, retention, retention. Retention simply means that if someone clicks on your video, you have to keep them watching in order for your content to perform its best.

There is an idea that continues to circulate that the human attention span has dipped below that of goldfish. The number put on it is eight seconds. This theory came from a study done in 2015, but a quick fact check leaves you wondering: Who comes up with this stuff anyway? A *BBC* article titled "Busting the attention span myth," written by Simon Maybin, exposes all the holes in the goldfish theory. It quotes Dr. Gemma Briggs, a psychology lecturer at Open University, who thinks the idea of an "average attention span" is pretty meaningless, because attention is task-dependent. She said, "How we apply our attention to different tasks depends very much about what the individual brings to that situation."

Whether human attention spans are getting shorter, there is one thing we probably all agree on: we live in a world full of distractions.

I woke up to 66 new text messages, 11 Slack messages, 4 voicemails, 128 Facebook notifications, and emails in the thousands. The cell phone has changed the way we manage the world, in both good ways and bad. We have the world at our fingertips and can be super informed and responsive. But this means that we're never disconnected. You might be in a meeting or having a real conversation with someone or at dinner or reading a book to unwind, and all the while your phone is vibrating or dinging with notifications. What's crazy is that sometimes your phone didn't actually do anything but you've been so conditioned to hear or feel it that you think it did. This happens to me all the time. These "phantom" rings or vibrations are common enough that the condition has been coined "phantom vibration syndrome" in an article written by Tim Locke on WebMD.

Respect Your Viewer's Commitment

With so many options vying for people's attention—Facebook, Instagram, Snapchat, TikTok, YouTube, Twitter, TV, news, text messages, hyperlinks, ads, and a million more—it's hard to cut through the noise as a content creator or business. So when a person chooses to click on your video, they are making a micro-commitment to say no to everything else and yes to your content. Respect that commitment. Even if the goldfish attention span rumor isn't true, you only have a short amount of time to hook a viewer and keep them watching so they don't leave for something more engaging. This is why it is so important to capitalize on the quality of your content. This is why you should spend more time making clickable videos that keep your audience engaged.

Engagement and Disengagement Triggers

When I look at a channel's analytics, I am looking for patterns among videos to see exactly where people stay engaged, get disengaged, and why. Then we know what to re-create to bring the AVD up. There is something that is extremely important to note here: every channel is different. Each has a different viewer and a different audience, so you can't compare graphs or expect similar patterns. Joe Rogan's YouTube channel has amazing Watch time because people watch his videos all the way through and then they watch more. Don't worry about Joe Rogan or keeping up with the digital Joneses. Just understand that the more people watch your content, the more data you'll have to work with.

Your goal is to find patterns and improve future content. Check your AVD in the video's first hour after upload, the first day, first week, first 30 days, and longer. Ask yourself what happened in the most engaging parts of your video. Did you use a call to action? Did you say or do something funny? Did you add an endscreen, suggest another video, or use an interesting filming or editing technique? Did you say a trigger phrase? Trigger phrases can reengage or disengage your audience, so make sure you do the former. Say things like, "Don't miss the bonus tip at the end of the video," not, "That's it for now."

I always ask Why as I'm analyzing. Why did people engage here, or Why did people leave? You either entertained them, brought value to them, or they got disconnected from you and they left. Continually connect and engage with people, and you'll grow. If you disconnect or disengage with people, your channel will stall or die. This is simple to explain but hard to implement. It takes time and patience to analyze. You have to commit to being dedicated to your data. Make yourself become more obsessed with the analysis process than you are about your view count. My students, clients, and I all deep dive into our analytics at the beginning of the month and again in the middle of the month. This allows us to see how our videos are performing and what we can do to identify patterns from engagement and disengagement triggers.

How to Read the Metrics

YouTube makes it easy to monitor everything that happens on your channel. For monitoring how long viewers stay on your video, there is a metric called average view duration, or AVD. If you have a 10-minute video and YouTube says your AVD is 2:30, that means on average your viewers watch 2 minutes and 30 seconds of your video. In case you're wondering, this is not a good AVD. The second metric to look at for retention is called average view percentage, or AVP. This is the average percentage of the video that people watch before leaving. So if your AVP says 63%, this means that on average your viewers watch 63% of the total video before leaving.

AVD and AVP do not match up perfectly because of the way people consume your video. For instance, I watch all videos at 2× speed. So on a 10:00 video, my AVD would be 5:00, but my AVP would be 100%. For long-term channel growth, these two metrics are key. The title and thumbnail will get you views, but AVD and AVP will get you Watch time and the right kind of audience. The longer you can keep viewers watching the more engaged they become, which turns into a loyal audience. The goal is to balance these two metrics and increase them together. YouTube provides multiple metrics to help with this. Take a look at Figures 16.1 through 16.3 as an explanation.

The "hockey stick" effect shown in Figure 16.1 happens when viewers come on and quickly leave. It's what happens when there is a disconnect between your title and thumbnail and what the viewer expected to see in the first 15 seconds of the video.

The "slow burn" effect shown in Figure 16.2 occurs when viewers come on and slowly leave over the length of the video. You want the line to be as flat as possible, so you're getting closer to the most optimal graph.

Figure 16.3 shows relative audience retention. This is a comparison between your video and other videos at similar lengths within

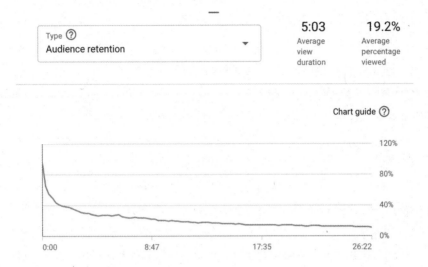

Figure 16.1 Hockey stick

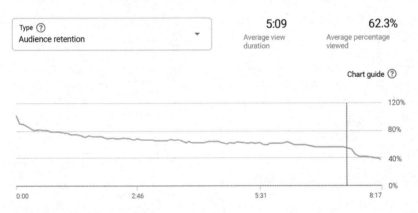

Figure 16.2 Slow burn

your niche. You're able to analyze how your video does next to others like it and analyze what grabs your audience's attention.

Look again at Figures 16.1 through 16.3. Notice the little "bumps." Sometimes the bump is subtle, but it's still there. The points of "rise" are where your content is performing well with your audience. Analyze what's going on in that point of the video so you can do more of it. Then look at the "fall" points, and analyze what was going on in your content to make people leave. In the Relative audience retention graph, look at the drop near the end of the video.

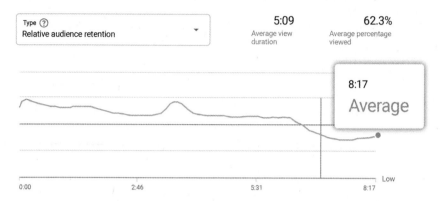

Figure 16.3 Relative audience retention

This was where the outro started, and people left. (The outro is the "wrapping it up" part where the person in the video can give a call to action and add endscreen elements like links to other videos or websites.) Do you have retention graphs that look like this? Find out why. Was there a trigger word or phrase that made them leave? Did you give them the promised payoff, so they had no reason to stay on to the end? What can you learn from this? These are the things you need to analyze when looking at this data. It will help you improve and connect with your viewer in future videos.

Average Views per Viewer

To know how well you connect with your viewers, you need to study your average views per viewer, or AVPV. The AVPV is the average number of times a viewer watched any video on your channel (in channel analytics) or this particular video (in video analytics). This metric gives you an in-depth understanding of whether your content really connects with your audience.

When you focus on the viewer and create entertaining content that connects, you can increase the amount of videos a viewer will watch over a period of time. Think of this like a TV series that people will watch over and over again, like *The Office* or *Friends*. Strive to make your videos so good that people become connected with you, the character. In the 28-day view in your channel metrics, look at how many videos you released in that time frame. Then notice how many videos the audience actually watched. You can even see what percentage was watched by a single viewer. This is a good way to understand who is watching on repeat and why.

The 50% Rule

On your retention graph, look for where 50% of the viewers leave the video, and go watch that part of your content to try to understand why people left at that point. You literally just lost half of your viewers. That's not a good thing. You want to retain as many people to the end of your video as possible. So if you can figure out where you lost half, you can figure out what you should change or avoid in your next video. Look for patterns to see if it's the same in similar videos.

Look at your endings and outros, too. Keep them short with the goal of getting the viewers to watch another video, preferably yours, fast. Have a verbal call to action and a link to another video that is relevant. You can put it as a description, a pinned comment, or as an endscreen element—it just depends on the content. Figure out what method works best in your niche by testing it yourself and by seeing what similar creators have done successfully, and do it. Most creators don't dig deep to figure out where and why they lost viewers. When you dig, you can fix what you did wrong for future videos so your content will have better retention. And better retention gives the AI more of what it needs to find similar viewers to push your content to. The more you understand your content, the better content strategy you'll have moving forward. When you can move that 50% mark to a further point on the content, YouTube will recommend your video more because it will have a higher AVD.

The 30% Rule

After you look at the 50% dropoff, go back and do it again at the spot where you lost the first 30% of your viewers. It's usually in the first 30 to 60 seconds of the video. Figure out what triggered this drop so you can see how to adjust based on the data. Again, implement tactics like the ones I listed in the 50% Rule above to reengage the audience. If you include a timestamp of the key points of your content, you'll see a more engaging graph where viewers went to those specific elements.

Maintain the Value Proposition

In June 2015, I went to a retreat with our team that was creating the Squatty Potty pooping unicorn ad. We got done writing the final script and thought it was amazing, but we realized that it was going to be much longer than any ads around at the time. Generally, most noninfomercial ads are 15 or 30 or 60 seconds. Long ads would have been a minute and a half. We were over four minutes with our ad. As we went through production, we thought we could cut down the content to be more digestible, but when we got to the edit, we knew we had something great just the way it was.

We asked a handful of trusted people in our field to watch our ad and give us feedback. These were people who were successful video marketers or content creators. Every single person told us the ad was too long. But in my observation of how humans interacted with content, I had seen that people would stay engaged if the value proposition remained. It had to provide value every step of the way, which I thought our long ad did. Other ads got to the point, did the sales pitch, and moved on. Our ad was different. We knew we could keep people watching because of the unique content even if they didn't plan on buying our product. Great content will keep people glued, goldfish be damned. If this wasn't true, the Netflix Binge wouldn't be a thing.

Jeffrey Harmon and I went back and forth for days about what we should do. We pulled his brother Daniel in to help us decide on a plan of action. Daniel was the creative director, and he also thought we had something good enough all the way through that we didn't need to cut anything out. Daniel did suggest that we punch it up with some sound design to make sure we kept engagement throughout, so he spent three days with the sound editor to get it just right. We also sped up the video 5% to get it just under four minutes. We launched the campaign. In the first full month, 92.6% of people who initially clicked on the video stayed and watched to the end. We had more shares, engagement, and sales because we kept people on the video, even as long as it was.

Jon Youshaei is product marketing manager of IGTV at Instagram and former head of creator product marketing at YouTube. He landed on Forbes 30 Under 30 for Marketing and Advertising in 2020. He also has a career-related cartoon series called *Every Vowel* (did you notice he has "every vowel" in his last name? So clever). I was talking with Jon about the concept of maintaining the value proposition when he said one of the most powerful things about finding the value for viewers in every video. He said, "You should spend hours of creative effort for every second viewers are consuming your content." You have to offer quality and value, because consumers' expectation for content is higher than it's ever been.

Message First

I need to give you a word of caution, however: don't be married to your content. Put effort and energy into the edit, and don't be afraid to cut what's not working. Become emotionally detached from your content so you can look at it objectively and it won't make you feel bad when you have to cut something. For example, in that same Squatty Potty ad, Jared Mecham, a YouTube creator and friend of mine, had a scene where he was an elf who cranked the conveyor belt to move the ice cream. Jeffrey said we needed to cut Jared's scene from the ad. I argued with him because the scene was so funny to me; it was a great pattern interrupt. Jeffrey stuck to his guns, and I'm glad he did, because when we cut the scene, the ad really was better. Jeffrey lives by the sales rule, "Message first, content second," and it's true 100% of the time. Never sacrifice the message of your content or your brand for the sake of comedy or anything else. Not only do you fall off brand, you'll likely lose engagement, too. For sales, it's always message first, content second. But for entertainment, it's content first, and message second.

Pattern Interrupts

Pay attention to natural breaks when you watch videos on You-Tube. Notice when you start to get bored or distracted. Do this on your own videos, too. That point where you start to disconnect is when you want to implement a pattern interrupt. A pattern interrupt resets your mind to behaviors or situations in the video. You can do this with new information that teases what's coming later in the video. You can interrupt by saying or doing something funny. You can add juxtapositions like a clip of something else, add a sound effect, or even simply clap your hands or change your voice. The most powerful technique is changing the camera's position with different angles or zoom. Look at your video's peaks and valleys in your metrics and add a pattern interrupt in the valleys to reengage.

MrBeast is one of the best I've seen at pattern interrupts. Go watch any recent video on his channel and you'll see what I mean. His brother CJ is just getting started with his own YouTube channel, so he wanted me to give feedback on a video he was working on. CJ did a great job and had great content, but it was a long video, and at one point, I said there was some disengagement. He needed a pattern interrupt. MrBeast smiled and said he had told CJ the exact same thing for the exact same moment in the video. Great YouTube minds think alike.

Mastering Engagement

The longer you create content and pay attention to the valleys, the better you'll get at recognizing when you need to reengage your audience. But you have to be an active viewer and pay attention to when people lose interest on your videos. Pay attention to what pulls you in and what pulls you out of a video so you know how to use engagement tactics. Always keep in mind your hook and your storytelling to keep your viewer interested. If you're not sure where to add pattern interrupts when you watch a video, the good news is that YouTube gives you the data to figure it out. Look at the data to know where you need reengagement.

YouTube's #1 goal is to predict what the viewers want to watch. When they watch more, it accrues more Watch time and AVD. People simply need to stay on YouTube as long as possible. So if your content is doing this, the algorithm rewards it. Creators always want to know exactly how much Watch time YouTube wants, but there's no hard and fast number. YouTube just wants more. More Watch time. More AVD. So create your content with More in mind.

Action Exercise

Task 1: Review your three best videos when it comes to AVD and AVP.

Task 2: Analyze using the 50% and 30% rules. How would you make your video different knowing what you know now?

Task 3: Notice the bumps and drop-offs in your videos and see if you can find a pattern.

Get the companion workbook and find more resources at www.ytformulabook.com.

17 Creating a YouTube Content Strategy

YouTube's growth statistics change so fast that any number I give in this book will likely be outdated by the time you read it. According to Omnicore, current statistics show there are over 31 million YouTube channels. Not videos, channels. (The number of videos runs well over five billion.) Obviously, these millions of channels have topics or themes that range in variety across the board. There's a channel for anything you could imagine. By the millions. The sheer volume of channels, in addition to their inherent diversity, makes a one-size-fits-all content strategy sound absurd and impossible.

In Chapter 13, you learned general strategies for storytelling and patterns of narration that work comprehensively. The same is true for your strategy to make that content: there are things that work, no matter what type of channel you have. If you want your channel to grow, your content strategy should be your top priority. As we talk about content strategy in this chapter, we'll talk about it in two parts:

1. Strategizing for your audience
2. Strategizing to leverage the algorithm

YouTube growth depends on your understanding how your content resonates with your audience so you can create more content like that. This includes things like your metadata and reengagement. Growth also depends on your understanding of data. YouTube gives us so much data that sometimes it can be overwhelming to decipher and utilize as we make more videos. All that data is a good thing (thanks, YouTube!), so try not to get frustrated. I'll help you understand how to use traffic sources and data relationships to your strategic advantage. The goal is to figure out how to combine part one strategy and part two strategy just right, because when you do, your content will be primed for massive growth.

Part One: Strategize for Your Audience

As the creator, your first job is to provide a smooth and predictable experience for your current viewers and subscribers. You want to create a feeling of security for them, a "same time, same place, same channel" vibe. They need to recognize that it's you every time they come back. Be predictable. A consistent channel is a successful channel. Your viewers should be waiting and anticipating your next video because they know your schedule and expect it from you.

Create

YouTubers always want to know what content to create. They often ask me questions like:

- What video content should I make to best engage my audience?
- How do I make content that is different or unique?
- How can I stand out while staying loyal to my brand?
- How can I create online content at scale?

Here's the hard truth about scale: There's no way a creator/brand can create all the content needed to feed consumers' voracious appetite for video, especially on mobile devices. There isn't enough time, money, or resources. The trick is to create content gradually and build an engaging library over time. That might sound daunting with a traditional production mind-set as a reference point. But to produce at scale requires rethinking that huge production process. You don't need a fancy production set and equipment to scale; in fact, you literally can use your smartphone and be successful if you create engaging content.

Create might just mean you make an entertaining video that gets people's attention. But as you consider what your audience cares about and will engage with, think about the micro-moments they might be experiencing. Micro-moments happen when people turn to their devices to find answers, discover new things, make decisions, be entertained, or buy something. You can find more on Google's guide to micro-moments in Chapter 11. Make content that fulfills your viewers' needs, whether that means you entertain, inform, inspire, answer, and/or sell. Also, think of using video for storymaking that viewers become a part of rather than telling them a story without engagement.

With millions of channels, the call to "be different" might sound like a joke. It can feel crowded on YouTube, especially in popular content genres. Even so, there is always a way to make your own spin and figure out how to be different. For example, Kristen from the cooking channel *Six Sisters Stuff* knows cooking channels are a dime a dozen. There are probably millions of cooking channels. Kristen carved out her own place among them by focusing on easy Instant Pot recipes, and she has a really successful channel. I've heard every excuse for why your genre is difficult to breach. Someone else has found a way to be different in your genre, so put on your thinking cap and do some research. You'll figure it out.

But how do you stand out without changing who you are? The short answer is to be creative without sacrificing consistency. There is always a way to stand out! Yes, some creators have chosen to sell out in order to stand out, but you do not have to change who you are.

Collaborate

Dan Markham wanted to make a soccer video for his channel *What's Inside*. He bought three World Cup soccer balls from different years off of eBay to compare them. He knew the video would be more interesting if he had a real soccer player kicking the balls around, so he collaborated with another YouTube creator, Garrett Gee, who was a former collegiate soccer player. Garrett had lived in Russia, which happened to be where the World Cup was occurring that year, so he also had some interesting, relevant cultural input. At the end of the video, Dan cut open the balls per his trademark question "What's Inside?" He released this video in conjunction with the World Cup tournament, and both his and Garrett's channels got a ton of views. In fact, it pushed Garrett's featured video to get over a million views—his first million-viewed video ever.

Collaborations can push traffic that wouldn't have found your channel otherwise. We talk in depth about the benefits of collaborations in Chapter 19, so don't miss it. I give several examples with Brooklyn and Bailey McKnight, identical twin vloggers who do some successful collaborations with BYUtv's *Studio C* sketch comedy channel. Their paired strategy got *Studio C*, a small channel at the time, a lot of exposure and a lot of money. Of course, it helped Brooklyn and Bailey, too. Your Watch time, subscribers, and revenue also can get a major boost from collabs, so make a place for them in your content strategy.

Curate

To have a successful YouTube channel, you have to be organized. Make a content schedule. Define your target audience. Provide a value proposition and deliver on the promises in your title, thumbnail, and in the video itself. Be consistent with your brand in everything that you do. But don't drown in your schedules and checklists; keep it simple. Prioritize the things that should never get neglected, but allow for wiggle room when things come up, because they will.

Make a content calendar. It will serve as a week-to-week outline of:

- What content you're posting and when
- Which audiences you're targeting
- Which social channels you're hitting

I can't tell you exactly what your calendar should look like, because of course, every channel is different and every creator is different. But I did make a generic content schedule template that you can customize for your channel. Go to www.ytformulabook.com to get the template and additional training. You should know what your optimal schedule is based on your own channel's impressions, click-through rate, and views. One of my clients posts two videos a day, while another posts one a week. NASA scientist turned YouTuber Mark Rober posts one video a month, and his channel is doing awesome because his monthly content is awesome. Use the tools available to help you understand where your traffic is coming from to help you strategize. As a friendly reminder, always prioritize your strategy for recommendation, since it accounts for three-quarters of all YouTube traffic.

Don't overthink your YouTube content marketing strategy. With brand marketers, there is often a disconnect in their content strategy between knowing what content to make and how to make it. Don't waste time overcomplicating it: video is what consumers want on mobile, so optimize for mobile. Create videos that fall in the crossover between your passion and what your audience wants.

Know Where Viewers Are Coming From

YouTube looks very closely at the audience, so what does that tell you? That you should too. Do you think about where your viewers are coming from before you make a new video? Every one of you—even those with millions of subscribers—needs to understand your audience better. You need to know how they are watching your content. My rule of thumb is to optimize for mobile first, because it's usually where people are watching YouTube. Mobile traffic accounts for 70% of daily Watch time. Most creators don't take this into consideration when creating new videos. Go to your traffic sources and see where people are coming from to watch your stuff. If they are coming from mobile, and you are on your computer, switch to your mobile device and go back to experience your video the way your viewer experiences it. When you look at YouTube from your viewer's perspective, you'll be super sensitive to how they behave and what works for them. You'll likely notice things you can change to optimize their viewing experience when you watch how they watch.

As you create new videos, know which traffic source you are optimizing that particular video for, because it will change how you create and what you do with it. There are times when it works well to reach an audience that is finding videos on their own from a search bar. There are times when it works well to reach them in a collaboration with another creator or business. And there are times when it works well to reach them from your video description, cards and endscreens, playlists, community posts, and stories. However, 75% of all video views on YouTube come from YouTube's recommendations. The algorithm follows the audience to know what to recommend. If you can develop an engaged audience, the algorithm will pay attention to what is keeping them engaged and will recommend more of your content—if you've done the work to connect your content. Let's take a closer look at how to connect your content by strategizing for the algorithm.

Part Two: Strategize to Leverage the Algorithm

As you've learned many times in this book, the AI observes a viewer's patterns and behaviors to predict what they will want to watch. Because of this, your second order of business is to help the AI understand which content goes together. If you don't do part one, strategizing for your audience first, part two will never happen.

The consistency tactics you implement keep the audience happy but also give the AI consistent information it will recognize and group with similar content. The AI will recommend your linked videos in conjunction with each other and to the viewers it has found that would be interested in them. When you've created a pattern with those videos or you've created a series of videos, you've made YouTube's job of recognizing your content a lot easier. The AI loves predictable data patterns and related content. A good place to look first when strategizing for the algorithm is at your Search traffic.

Search

Search has been around since the advent of the Internet. "Search" means people go to the search bar of a website and type in a query. They don't simply type in a keyword or two, they are good at getting laser specific to ask exactly what they need an answer to. For example, someone might search, "How do you use a septic tank with a pool?" instead of searching, "pool septic tank." With every client of mine, I have them make a list of all the frequently asked questions that people ask in their niche or business. You need to do this, too. Don't worry about arranging the list in any particular order yet, just write down your own FAQs. This works for all businesses and YouTube creators.

Once, I had a client who wrote a list of about 40 of their most frequently asked questions. Forty sounds like a great list, right? Well, to get the most out of this exercise, you have to go way beyond FAQs. So after he had his 40 down, I had him think about every possible question he had ever been asked and had him add them all to his list. He ended up with more than 250 questions. Now you can sort your list according to which questions come up the most. Then you are ready to create content that answers all of the questions your potential viewer might ask in Search, giving the most asked questions priority. If your content pops up at the top of the results, viewers are more likely to click on your video, and you're more likely to have gained another viewer. You also need to consider that YouTube is owned by Google, so your content needs to show up in a Google search as well, because your traffic could come from there.

Next, I want you to make a list of questions that people *should* ask you. These are your SAQs, or should ask questions. Your SAQs and FAQs should interweave. How? Let's say you create a video to answer one of your frequently asked questions. In this video, you also should bring in an SAQ that relates to the FAQ at hand. This strategy satisfies your viewer's curiosity, gives them even more than what they came for, and still provides a smooth viewing experience that is digestible and brings value. Further, another great strategy to use here is to create another video (or two or three . . .) with the same line of questions in mind. Then you can recommend they watch it next, either verbally or with an end card or both. There's also a higher probability that YouTube might put it in the Suggested videos. This is a great way to keep the viewer engaged and watching more of your content or buying your products. When you do this, the AI is very happy and will reward you by pushing your content to other similar viewers.

I like to look at Search as an entry point into your world of content. Most creators and YouTube educators focus on Search and ranking, but for me, I look at it as a funnel for bringing in additional traffic. When you get the initial traffic on your content, you can turn it into consumption traffic, meaning you've converted your first-time viewers into repeat consumers of your content. Matthew Patrick, or "MatPat," is a YouTube creator who has successful channels in a variety of genres. He uses Search well as a funnel. For example, if you search "Do video games cause violence?" MatPat's video "Do Video Games Cause Violence? It's Complicated" tops results including serious science and news channels. But what's important is what you see after you click on his video—several of his other videos suggested to watch next. He now has a new viewer consuming his content. He's done the same thing with other highly searchable FAQs in his genre, like, "How does the Force work?" and "What is a Yoshi?" (These actually are some of the most important questions you could ever ask. Anything to do with Star Wars or Super Mario Bros. should be at the top of your need-to-know list.)

This works for businesses, too. Gillette, the razor company, made a video to answer the question, "How to shave." They made a whole series of videos to provide solutions for any kind of shaving issue a person might have. These range in topic from learning the basics to more advanced shaving techniques. Answer people's questions, and make sure you've connected your content so it's right there for the viewers to click on next. Even when your demographic shifts, you'll have a new generation and audience coming in and discovering you because they are searching for an answer to a question.

Recommended/Suggested

YouTube's recommendations account for 75% of all views on YouTube. Did you notice that this is not the first time I used this statistic?

(Cue: pay attention.) You want your content to be where 75% of the views are coming from. YouTube recommends videos in three places: on Browse through the Homepage and Subscriptions, the Trending tab, and Suggested videos in conjunction with a video that is currently playing and what's up next. The Trending tab features geo-specific popular topics that have a broader appeal. Take different approaches to your content strategy based on which place your traffic is coming from. Each traffic source's algorithm has indicators to know when content is good and when it is bad. YouTube takes the videos that indicate "good" and puts them where they will likely get the best response.

Some creators think releasing as much content as possible gives them the best opportunity to be seen. While this can be true in certain situations, like for kids' content and gaming content, it can be detrimental to your own content. Multiple uploads too close together means the algorithm has to pick one to push. And when it picks one, what happens to the other? Traffic jam. One gets through, but the rest come to a complete stop.

The algorithm specific to Browse traffic keeps track of what the viewer clicks on and what they don't click on so it knows what to put on Browse next time you come to YouTube. I recently clicked on a video that was recommended to me on Browse because the thumbnail was so good that I had to click to satisfy my curiosity (plus, my sons Bridger and Thatcher had already asked me to watch it). It was called "Work Stories (sooubway)" by channel *TheOdd1sOut*. After watching the video, I binge-watched that creator's content. So, the next time I logged on to YouTube, who do you think was recommended to me again? Of course, it was that same channel, and I hadn't even subscribed to them. (Don't worry, James, I am a subscriber now.)

Browse on YouTube's Homepage is the fastest way to get views on your video. This is very appealing for obvious reasons; you don't have to wait to get a lot of views and momentum. If this is your content strategy, be warned that it's easy to get burned out. If you are Browse-focused first, you need to have a Suggested feed strategy, too. Most daily vloggers don't have a Suggested video strategy. Over years of observing and working with a lot of daily vloggers, I've learned that the most optimal Browse schedule is three times a week. A lot of vloggers think that in order to make it big, they have to release a new video every single day. It makes sense that anyone risks burnout at that rate. Lucky for them, it actually works better to have some natural breathing room between uploads, both for the viewer to have time to watch the content, and for YouTube to figure out who to push the content to and what videos to suggest after that video.

Rather than trudging through an endless daily grind, figure out how to tie your content together. Do a series of videos so people have to watch all to get a complete story. Or you can have a recurring theme. This will help you build a bigger video library, which increases your opportunity for growth on the golden Suggested feed. Then you won't be relying solely on your Browse traffic to carry the weight of your entire channel.

Let me tell you about a channel whose content strategy includes having a huge library of videos. World Wrestling Entertainment, or WWE, gets between 1.4 and 1.5 billion video views *a month*. They've been known to upload 24 to 50 videos in one day. This insane schedule would overwhelm Browse, but it works great on Suggested. If your channel has close to 50,000 videos (which WWE does!) YouTube will put your new content on the Suggested feed. Don't forget to tie in this new content with your old content, though. New content that has a data relationship to your old videos gives them a freshness factor to be suggested again, even though they might be years old. The longer your content is on YouTube, the more views you'll accumulate with this strategy. This can be extremely profitable.

We had a channel that we uploaded three times a week until we had 1,200 videos in our library. We took a break from uploading for a whole month but continued to get views and revenue from our library simply because there was a lot of content there to continue to be recommended.

Amy Wiley, WWE's senior director of YouTube operations, gave great insights into their company's strategies. Amy said that if there is another channel sending views to your channel's videos, it's helpful for you to know about that relationship. You can go to their video and see why your content is being recommended alongside it. You might see topics you're not covering that are working well with your brand and that you should be covering. Or you could have a potential collaboration with that video's creator. One month, WWE gained two million views that came from a single video from another channel. "This shows the power of the algorithm," Amy said, "and the power of being a part of the YouTube ecosystem." She nailed it on the head. You have to figure out how to get the algorithm to see your content as a contributing part of the system.

WWE is really good at paying attention to what's working on other channels and modifying their own content to match. For example, they knew arm wrestling had been popular in the past but wouldn't have considered it a current trend. Then they noticed that arm wrestling was seeing a resurgence. So WWE put together some arm wrestling clips in a new video. It got more than 23 million views. This topic hadn't been on their radar, but by paying attention to what others in the space were doing, they added the same type of content to their own strategy, and it really paid off.

Data Relationships

Every new video uploaded to YouTube gets a digital ID. We talked about the Content ID system in Chapter 2 and why it was necessary. The ID is like a digital fingerprint. In addition, when you create a playlist, that playlist also gets an ID, as does adding a video to an existing playlist. When you add videos together in a playlist, this tactic to get viewers to binge watch works really well because viewers watch them in a specific order. Also, YouTube notices the IDs that are grouped together. So if someone is watching a video that was tied in from a playlist, then there is a higher probability that the algorithm will plug in that related video to the recommendation feed. This doesn't happen every time; it's based on what the viewer does next. But if you have a small percentage of viewers watching the videos sequentially in the playlist, it increases the chance that YouTube will recommend that content based on your data relationship and viewing relationship.

Weave your content strategy so your videos are related by topic and nature. This will increase the likelihood of getting your videos suggested. This is where the rubber meets the road, and your view count will take off. It's what I call the "Goldilocks Zone." You remember how that girl with the golden hair liked everything "just right" at the house of the three bears? It's like that. A content strategy that caters to both the audience and the algorithm "just right" gives your videos the best chance at being suggested and seen and to really take off.

Another great way to create a data relationship among your videos is to put your top performing videos as endscreens and cards on related videos. Even if only 2% of viewers watch to the end of the video and click on your recommendation, that's a lot from the algorithm's point of view. YouTube can go out and find other viewers that have that same type of data relationship of that 2% and recommend that video to those potential viewers. If these potential viewers engage and like your content, then YouTube will continue to recommend not just this video, but more content from your library. That little percentage stimulates another look at your content from the AI, which will continue to find viewers with similar viewing patterns and push your content to them, too. Every little push helps.

Video Description

The video's description is often overlooked because it doesn't produce quick or massive results, but using this feature can help you in smaller ways. The description, besides giving good keywords for Search and a summary of the video, also tie in links to other videos and playlists. Most people do not open the description, but some do. Again, even though this percentage is small, any additional activity from the viewer on that video creates a viewing pattern for the AI to connect the data. As they click the next video, that relationship between the two videos deepens. It also creates more Watch time on your channel. Every time those few people click on your description links and watch more, YouTube rewards you. Additionally, the viewers who are clicking in the description are usually the most active viewers on your content, so YouTube watches their behavior more closely than less active viewers.

Playlists and YouTube Mix

Playlists can be powerful tools for retaining and engaging active viewers. As a creator, you have the ability to create playlists for your subscribers and viewers to watch back to back. Viewers can watch your videos almost uninterrupted because the next up video isn't one that the YouTube AI is suggesting but the next video you have added to that playlist. For those of you born before 1985 (or if you've seen the movie *Guardians of the Galaxy*), it's like making a mixtape, where you record your own selection of songs onto one cassette. This is a great way to improve the viewing experience by creating content that was meant to be consumed in the order you decided. It helps the viewer know which video in that series of videos to watch next. I love playlists, because playlist viewers bring in the most Watch time on your channel. I've created many successful content strategies based on playlists. Playlists emphasize the viewer's journey with your content.

Let me give you a quick example of how YouTube content could be made into a series. Let's say I had a technology review channel where I featured the newest tech products. I could create a video of the "rumors and speculations" of a new iPhone coming out and talk about its features and projections. This would be a great video for this type of channel. Then I could make a video about the announcement of the new iPhone. For another video, I could show myself using the new phone for the first time after it came out. Another video could be a teardown of the phone with its components exposed. Still another could be a video listing all the pros and cons of the phone after using it for 15 days (not the teardown phone but a second phone). You could add all these videos in a playlist for a new viewer to consume in one setting like it was meant to be seen back to back to back.

Most creators don't use playlists this way; they throw videos on YouTube and add a title with no rhyme or reason. By creating a playlist meant to be watched sequentially and tailored to your viewer, you are creating an environment that will get you your best Watch time. Which means the AI will usually start recommending those videos to be watched in that order as the next up.

The discovery team at YouTube noticed playlist viewing data and decided to make their own ultimate mixtape or playlists for viewers. The AI does a great job predicting videos for viewers to watch, so they tested a new feature called YouTube Mix. YouTube Mix is a non-stop playlist that the AI tailors to the viewer. This feature has created amazing Watch time for YouTube.

Community Posts and Stories

The YouTube Community tab is a feature designed to help creators engage with their audience outside of the videos that they upload to their channel. Types of Community posts include polls, text-based posts, and images. It's a great place to promote older videos or merchandise or other products that you sell. It's also a great place to ask your audience about ideas for different videos. Creators have found these ideas to be super helpful at catering to what their viewers want that they wouldn't have thought of on their own for a new video idea. Community posts can be found in the tabs on a channel, usually between Playlists and Channels or Store, if you have one. (If you don't see one, it's because YouTube considers your content geared toward kids, or you haven't met the subscriber minimum.)

Stories are short, mobile-only videos that allow you to connect with your audience more casually on the go. Stories expire after seven days. They appear on the Homepage and look like stories on other platforms, with the rolling view moving right as new stories come in and are placed at the left of the scroll. The Story feed is like putting up a banner or a billboard for your brand.

YouTube pushes your Community posts and Stories to both subscribers and nonsubscribers. I've seen creators grow massively from using these features. These are subscriber-based recommendations, meaning the algorithm's goal is to push Community posts and Stories to viewers who have seen your content but are not subscribed to your channel. It also means the algorithm will push posts and Stories to potential new viewers and subscribers.

Tent Poles

To accelerate visibility, use a "tent pole" strategy. Think about how a tent pole holds up its side of the tent. There is a "peaking" effect at the top of the tent. The peak represents an event or something relevant to a group's general knowledge. This could be a holiday. It could be the opening day of a big movie. It could be the Super Bowl or the World Cup in soccer. Align your video topic and content around one of these tent pole peaks to accelerate your channel's visibility in conjunction with that thing.

For example, when I was working with Jared Shores, the cocreator and director of BYUtv's comedy sketch channel, *Studio C*, we looked at a calendar and brainstormed what regional, national, and worldwide events would be happening that year. The other cocreator, Matt Meese, was also a head writer and actor. He brought up an idea he had always wanted to make into a video with a soccer match. So we looked at when the men's NCAA soccer championship landed on the calendar, mid-November, and we set our sights on a video release. We used two big schools, Yale and North Carolina, as opponents in our video. Matt wrote an amazing script. The goalie, Matt as "Scott Sterling," had the unfortunate role of taking a soccer ball to the face over and over (with the help of some great VFX, good editing, a soft soccer ball, and good make-up).

Three days before the NCAA championship, we did a subreddit post with the Scott Sterling video. Soccer fans who saw it knew it was fake, but they loved it because it was so funny. And because it was soccer. This exposure and momentum helped the post land on the front page of Reddit, and it was shared and reshared elsewhere online. We got 12 million views in less than 24 hours. It catapulted the small channel. Within a few days, a rumor started circulating that the goalie had died from his injuries. Of course, this wasn't true, but it continued to generate buzz for us. This video release strategy hit a double-whammy tent pole: we timed it around the US national soccer championship, but also we appealed to a worldwide soccer audience.

Let's go back to the soccer video we talked about in collaborations. Dan Markham used the help of his soccer friend Garrett to make an interesting video about World Cup soccer balls, which he eventually cut open at the end of the video. You already know that this was a successful collaboration strategy, but Dan's content release strategy is very important for us to talk about as a tent pole as well. Dan had filmed this video in January, but the World Cup tournament wasn't until June, so he just let the video sit.

In June, he started watching Google's trending page. On day one of the tournament, "World Cup" was really trending, so Dan knew it was finally time to release his soccer video. He uploaded at 1 a.m. where he lives in the States, which was good timing internationally. The US soccer team didn't even make it to the tournament that year, but the international hype around the event generated a lot of views for Dan's video, and for Garrett Gee, the YouTube creator he had collaborated with.

Tent poling is a really smart tactic to add to your overall content strategy, especially because the content is usually evergreen, meaning that that content gets a boost every time that event comes back around. Mine your video archives to see what content you've already made that might fit into a tent pole strategy, then make new content with data relationships connecting these videos. I promise it works for topics besides soccer!

Video Buckets

Once you get a good handle on your basic content strategies, you're ready to take it to the next level with something I call video buckets or content buckets. Buckets are an associated grouping for related videos, keywords, behaviors, and audience characteristics. Each bucket is a similar topic or theme that appeals to your target audience. Consider "bucket" the equivalent of "category."

Imagine you have a bucket. You put a treat in the bucket and set it on your front porch. Meanwhile, your neighbor has set out a row of different colored buckets, and each bucket contains a different treat. Every day, a group of neighborhood kids walks down your street and chooses which house's treat stash to raid. Which house do you think they will choose? Which house would you choose? It's safe to assume we all would choose the house with lots of different kinds of treats.

People who frequent the YouTube neighborhood also want variety. As much as they might love that one amazing video on your channel, they don't want to watch it every time they visit. This is why having a variety of "buckets" to draw from matters. If all you ever offer is cookie videos, your viewers will stop coming back for more cookies, because sometimes they want a donut or a candy bar or a soda pop.

When I was a kid, I loved a TV show called *The A-Team*. The A-Team was a group of Vietnam veterans who were framed for a crime they didn't commit. So they escaped jail and decided to become mercenaries who helped the good guys get away from the bad guys. In every episode the plot was the same: good guys get into trouble, the A-Team shows up, chaos ensues when the Team hatches a crazy plan, and in the end, somehow they accomplish the impossible. At the end of every episode, a character named Colonel Smith would say, "I love it when a plan comes together." The episode always had a crazy build-up and they were always modifying the plan.

Somewhere in the middle of season two, I got bored with the series. Their ratings plummeted because everybody got bored with it. Every single episode felt the same. It was basically the exact same plotline with different characters. You might have a really great idea for content, but if you overuse it, it will stop working, just like *The A-Team*.

Once you understand which niche you fall under, you can break down your own channel with video buckets. Breaking down your content into categories helps your audience know exactly what they're going to get with a particular type of video, and they get their payoff when you deliver. Then you can wash, rinse, and repeat on each type of video. When you do it right, you'll have a low churn rate, meaning you'll lose fewer viewers. To be clear, having categories or buckets is not the same as having a variety channel. Variety channels don't work. Buckets are categories that are related, not random. Narrow down your niche, but don't be so narrow that the viewers lose interest because you have no variety.

For example, MrBeast is a video buckets master. He has a very high standard for himself as he creates new content. If it doesn't fit into one of his video buckets, he won't put it on his channel, even if it's a great video. His buckets include categories like extreme challenges, last to _____ wins, would you rather, and 24-hour challenges. MrBeast has a very low churn rate.

When you create content buckets, you'll realize it is a great way to systematize programming for your channel. It will help you schedule so you'll know exactly what you're going to do ahead of time. Create editorial calendars, schedule content, and craft copy with ease knowing the various topics you should be hitting. This will save you so much time and stress.

With that said, don't be too rigid with your buckets. Topics can ebb and flow depending on what's currently working. If something stops working, experiment with a new bucket. You should be experimenting always anyway. If you don't, you'll be stagnant and eventually canceled like the A-Team.

How to Choose Your Buckets

Choosing your channel's buckets is really fun. Go to your videos and look for patterns among them. Especially consider your top performing videos and make sure you create a bucket for them. Group similar videos and title each bucket. Refer to your FAQ/SAQ lists you made earlier in this chapter to make sure you cover the important topics.

I introduced you to Kristina Smallhorn in Chapter 10. She is a Louisiana realtor who came to me for help with her channel. She had been creating content based off of keywords and SEO. I helped her categorize her top performing videos into buckets and told her to create more content for those buckets. She connected metadata and video structure, and before long, her channel took off. A good bucket strategy was all she needed to take her channel to the next level. Check out her spikes over time after implementing buckets, shown in Figures 17.1 through 17.5.

Figure 17.1 28 days

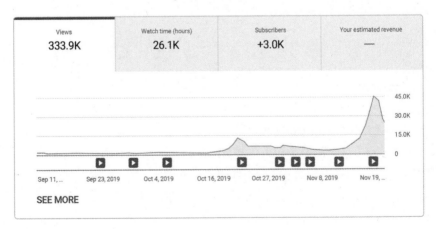

Figure 17.2 90 days

People watched your videos 1,180,400 times during the dates you selected

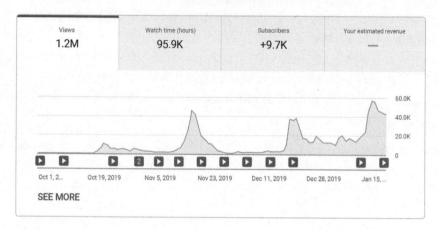

Figure 17.3 4 months

People watched your videos 2,157,897 times during the dates you selected

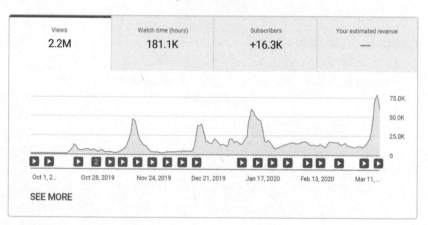

Figure 17.4 6 months

Look at the first tiny bump in Figure 17.5 (the arrow is pointing to it). It's the spike that looked so big in Figure 17.1. You can see that the buckets strategy perpetually grows over time and as you release more videos in that bucket.

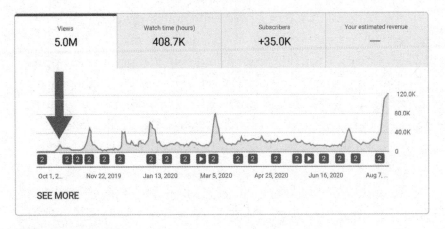

Figure 17.5 1 year

How to Create Buckets

How do you put a video "in the bucket?" Once you've labeled your content buckets by topic or theme, make sure the videos that are grouped together have similar metadata. They should have similar titles, keywords, descriptions, and tags. The videos should be similarly structured as well. The framework, pacing, and editing should follow specific patterns.

YouTube used to only recommend exact content matches as suggestions. But they found that in order to keep the viewer on YouTube longer, they needed to have the viewer follow one of three paths: rabbit holes, similar videos, and personalized for viewer. The "rabbit hole" recommendations take a viewer deeper and deeper into one topic or one type of video. Similar video recommendations include videos that would be highly related but not super related like the rabbit-hole type. These videos have a natural relationship, whether it's viewing data or topic data. The personalized for viewer type takes something the viewer has engaged with that's not related to what they're currently watching but they're likely to watch based on their past viewing and subscribing patterns.

You want your content related enough that the algorithm will recommend it together. The connections you've made in advance make it easier for YouTube to find your stuff and put it where people will see it. If you only have one bucket, your content might take off, but eventually it will slow down because there's not enough there for YouTube to recommend in a variety of ways. People like all kinds of treats, not just cookies every day, remember? This is why buckets work so well on YouTube today.

To use a real-world example, let's look at one of my clients Devin Stone's channel *Legal Eagle*. Devin is a civil litigation lawyer in Washington, D.C., whose mission is to help explain everyday laws and legal issues to regular people. He knows his video buckets, and when he makes a new video, it always fits into one of them. Devin's buckets include the following and are shown in Figure 17.6:

REAL LAWYER REACTS TO   REAL LIFE, REAL LAW REVIEWS   LAWS BROKEN

LAW 101!   EXPERIMENTAL

Figure 17.6 *Legal Eagle* **buckets**

- Real Lawyer Reacts to _____
- Real Life, Real Law Reviews
- Real Lawyer Responds to ____
- Laws Broken ____
- Law 101!
- Real Lawyer Versus
- Experimental

Go watch one of Devin's videos and scroll down the Suggested feed. The Up next video is likely another *Legal Eagle* video that fits into the same category as the one playing. The second position might be occupied by a playlist of his channel library, and the third position might be another video of his within the current bucket. In Chapter 11, I explained how Devin figured out how to reach a broader audience than just the law students he had been creating for. Buckets changed the game for him, helping him reach more of a mainstream audience.

It's easier to brainstorm new ideas when you've written your content down and broken it into categories. When the lists are right in front of you, new content ideas strike more easily. You can also see where you might be lacking content. You can create videos that fit into multiple buckets, too. In fact, it's awesome to have some crossover. MrBeast does a lot of "challenge" type videos that easily hit more than one bucket. This works great because you can use a keyword to connect the buckets to get more of your content recommended.

It makes sense that having a well-rounded content strategy would keep your viewers' attention better than a one-themed wonder, doesn't it? In creating content bucket lists, you might find that your channel isn't as varied as you thought it was. You might come to the realization that you do only have one type of bucket. You might find that you have a bucket that's not really working. You might see an obvious oversight and be able to create a new bucket that gives your channel the boost you've been wanting.

The purpose of all of this is to make your content creation strategy easier and more effective at getting and keeping viewers. A good bucket strategy provides all the elements that retain engaged viewers. Engaged viewers become part of your community that is loyal to your brand. What more could a YouTube content creator want?

Keep It Simple

It's easy for YouTube creators to get stuck in a content strategy rut. Yes, it's important to plan, and it's important to implement tactics to utilize your data, but sometimes the answer is to keep it simple. As long as you remember to strategize for your audience and strategize for the algorithm, you'll be fine. Optimize for mobile and create data relationships among your content so YouTube will keep recommending you to the viewers. And don't forget the first part of this chapter's title: Create. Keeping your content fresh and interesting is always the best strategy.

Action Exercise

Task 1: Look at your video library and group similar videos together.

Task 2: Label these groups of videos by topic. These have become your buckets.

Task 3: Analyze your videos in your buckets using the Four Ws (Chapter 14's Action Exercise).

Task 4: Analyze your titles and thumbnails in each bucket by looking for patterns among the ones with the highest CTR and AVD.

Task 5: Plan, create, and upload a new video that will fall into one of your buckets.

Get the companion workbook and find more resources at www.ytformulabook.com.

18 Building a Community around Your Content

Over the course of my career I have looked at *a lot* of content online, some of it incredibly amazing and some painfully awful. I wanted to know why some content performed well and some fell flat on its face. It taught me to have an eye for patterns that worked. I began noticing that content from certain brands or people worked almost every time, so I looked deeper to figure out why. Of course, it mattered that videos were made well, but I discovered that what mattered most had less to do with the video itself and more with who was watching it. When content was built around a loyal community, it consistently performed better than content that wasn't.

Community is everything. Humans need each other; we yearn to connect in all kinds of ways. We need to feel a sense of belonging. We want to be a part of something bigger than ourselves. We like to be a part of a team, become a member of a band's groupies, meet with like-minded people at conventions and athletic events, and talk to others like us in groups online. These communities can be as broad as nations or as specific as men who like My Little Pony and call themselves "Bronies" (see Chapter 12 for this fun story). You can belong to a small, local service club that helps families in need during the holidays, or you can belong to the million-strong Rotary International service organization. There is an amazing book about building an online community called *Superfans*. It was written by my friend Pat Flynn. Pat's insights are spot on when it comes to developing and activating an audience online. I strongly recommend that you read it after reading this book.

As YouTube creators, we need to establish our own communities, no matter what genre our channel fits into. But how? What does it take to build a community? When I work with clients, I always have them develop a plan to build their following before they create more content. Of course, you have to have content for them to share, but you want to make sure you create the content specifically for them. You can't do this if you don't know them first. How do you get to know them? In previous chapters, we talked about viewer personas and really understanding what makes your viewers tick. Once you know your audience, you are ready to convert them into a loyal community around your content.

To help explain, let me tell you how my partners and I built a loyal following around a TV series about the life of Jesus Christ called *The Chosen*, and it has nothing to do with religion. Every client I work with is required to read a book called *Primal Branding* that teaches the fundamentals of community building. It was written by Patrick Hanlon in 2006 and remains relevant through every change in trends and technology to date.

When my new partner Dallas Jenkins and I connected, one of the first things we talked about was building an audience. I taught Dallas the fundamentals of *Primal Branding* and encouraged him to read the book as we set out to create a culture. Dallas was completely on board with *Primal Branding* because community building would be the best way for us to raise money for the project—and we had a lot of money to raise. We talked about brands that have loyal communities, like Apple or Tesla, and even cult followings, like the Grateful Dead and Star Trek. We needed to follow their model of acquiring dedicated fans.

In the book, my good friend Patrick reminds us that humans simply want to feel included, recognized, and accepted. We want to belong somewhere, knowing there are people just like us. We need to feel like we are not alone. Communities offer this sense of belonging, but they come with their own set of rules, language, and identities. Dallas and I broke down who our avatar was, and we listed the steps it would take to build them into a loyal culture around *The Chosen*. We knew how important it was for us to find "our people." We needed a passionate social army who could see our vision and our mission and make it their own. Once they took ownership, we knew they would donate money, but more importantly, they would share and promote the project to others who would join and further the cause.

Let me take you step-by-step through *Primal Branding*'s fundamentals as we implemented them into our community-building strategy at *The Chosen*. There are seven fundamentals: Creation Story, Creed, Icons, Rituals, Sacred Words or Lexicon, Nonbelievers, and Leader. (Don't miss the bonus fundamentals, too.)

Creation Story

First, we needed to tell our Creation Story. Patrick tells us the Creation Story is the beginning of the brand narrative. As humans, we want to know who we are, where we came from, and where we are going. As communities, the same is true. People want to know the story behind companies, organizations, products, and even YouTube creators in order to choose whether to become a supporter. Just because it's "business" doesn't mean people don't want it to be personal. When people see the *story* part of his*tory*, it connects them to something in a meaningful way. It needs to be personal. Patrick calls the Creation Story a legacy, which invokes feelings of belonging to a family or group over generations. Legacy means leaving a lasting impact, while history is just about the facts in the past.

A well-known example of a brand with a good Creation Story is Apple Inc. Most people have heard about the two Steves—Jobs and Wozniak—who started the company at Jobs's childhood home. They sold a calculator and a Volkswagen van to be able to buy parts and build the first Apple computers on the market. They named the company Apple because Jobs had recently spent time at an apple orchard in Oregon. People love this Creation Story.

Dallas and I, along with our distribution partners Jeffrey Harmon and Neal Harmon, wrote down our thoughts, values, passions, and our Why. If we were passionate about this project, then there had to be people out there who could relate—people who had passion for the same thing. We knew that if they could see our vision, they would want to be a part of it as well. We already knew who our ideal avatar was, so we started making content that would spread our message and resonate with them. *The Chosen*'s pilot episode told the story of Jesus Christ's birth through the lens of a disabled shepherd. It was received with a lot of passion and excitement by people across the world. Now that they had seen our work, we needed to show them who we were and how we began to cover fundamental number one of building a loyal community.

Our Creation Story video, "The Story Behind The Chosen," was nine minutes long, and began with Dallas explaining a raw personal moment. He was sitting at home on the worst day of his professional career, despairing over the failure of his most recent film, when he had a clear recollection of a Bible story. It was the story of Jesus feeding 5,000 people with just a few loaves of bread and a handful of fishes. Dallas explained that he knew God wanted him to spread the message of Jesus Christ, but he didn't know what that would entail at the time. So he committed to bringing his loaves and fishes to the table, so to speak, and knew that God would multiply his effort.

Not long after this heart-to-heart with God (and his wife), Dallas received a direct message from a friend about his job being "to bring the bread and fish." When Dallas asked his friend why he had said that to him at that particular time, the friend responded, "It wasn't me . . . I felt led to tell you that right now." Right out of the gate, Dallas told our viewers how this Jesus series had been born, pulling them into our Why.

Next in our Creation Story, Jeffrey and Neal explained how they were drawn to the project and joined as *The Chosen*'s distribution company, VidAngel. Then I told how I joined the project and why it meant so much to me. I said that I felt like everything I had done in my life had been leading to this project, which I truly do believe. To be clear, the content in our Creation Story wasn't a script whose aim was to collect numbers; it was genuinely 100% how we all came to be passionate about *The Chosen*. We wanted the people to know that there weren't a bunch of stuffy businesspeople behind the show's curtain. We became a team because we were working toward the solitary goal of spreading our passion about the message.

Our project received more financial donations than we could have ever imagined. Because of this, we repurposed our Creation Story video to show our gratitude for the overwhelming response. This helped our followers see where the money was going and that it mattered to us. In the extended video, Dallas shared the results of the continuous crowdfunding campaign. He wrapped up the video by saying that our team had all brought our loaves and fishes, the investors had brought their loaves and fishes, and God would multiply our efforts. By the end of the video, the viewer knew exactly where we had come from, why we were here, and how they could become a part of our passionate community.

The Creation Story facilitates inclusion for your viewers, taking "my vision" to "our vision." They embrace your passion as their own, and they want to share it with their sphere of influence. Then their sphere of influence catches the passion and wants to share as well, and this is how content and community spreads.

Good brands always tell their Creation Story. When the people know where you came from and why, they are ready to know what you're all about, which is the second fundamental.

Creed

The second fundamental is your Creed. This is your belief system. If your values and principles aren't clear to you, how will they be clear enough that your followers know exactly why they want to belong to your community? They want to know what you believe in and why you've come together. According to Patrick, "The Creed is a central idea that everyone wants to be associated with." Some communities' first identifier is their Creed. Think of "Semper Fi" and the Marines, or Nike's "Just Do It." If you know the brand, you can recite the Creed. Our Creed at *The Chosen* has evolved as our community has grown. It began as a rally cry. We didn't want anyone in Hollywood controlling the creative process; we wanted to hold the gold to create the content because then we could create based on our own rules. This is why we needed crowdfunding to work.

As *The Chosen*'s community grew, our Creed changed based on feedback and interactions with our followers. In episode 7 of season one, Peter argues against Jesus's call to Matthew to leave everything and follow him. Matthew was Peter's tax collector, and he needed him to stay. He tells Jesus that Matthew's job as tax collector warranted him staying, that Matthew's situation is different from his own. Jesus's reply, "Get used to different," became a catchphrase that viewers really responded to. So it became a Creed. We have since printed it on merchandise. The same thing happened with the phrase, "Binge Jesus." Our viewers latched onto it, and we integrated it into our Creed and our products as well. Sometimes the Creed comes from the community and gets adopted by the brand or creator, and sometimes it comes from the top down. It's an organic process that you need to be aware of as you pull your community together.

Icons

Now that they know where you came from and what you stand for, they need to know how to identify you, which brings us to the third fundamental: Icons. Icons are any representation of your brand. Patrick's explanation is simple, but perfect: "Icons remind us that we are in this place, not that place." Like if you walk into a shoe and apparel store and see a "swoosh" printed on every product—I don't even have to tell you you're in a Nike store, not Adidas or Reebok. The Icon is so ingrained in us. Mickey Mouse ears = Disney. The American flag = the United States, patriotism, pride, and more. As you can see from this last example, Icons can be associated with more than a place or a company; they can stand for values, visions, and feelings. Shapes, sounds, smells, and tastes all can be iconic as well. Patrick also says that Icons are "quick concentrations of meaning that we instantly identify with and feel something about." They can be logos, packaging, product design, experiences, and environments.

For *The Chosen*, the "Get used to different" phrase reflects one of our Icons. In our opening credits, there is a sea of gray fish swimming in a circle, while only a few turquoise fish swim against the current in the opposite direction. This symbolizes Jesus Christ and his followers moving in a completely different direction than the majority of the crowd. This Icon can be found on our products as well. When our community sees it, they identify with it and associate it with the community.

Dallas and I had discovered that we were both big fans of a graphic designer-filmmaker named Saul Bass who had created iconic designs for many popular film posters and movie title sequences. His work was popularized in the 1950s, spanned four decades, and included works like *Psycho*, *The Man with the Golden Arm*, *The Shining*, *West Side Story*, and *Big*. We wanted to capture a Saul Bass essence for our title sequence, and Eric Fowles at Voltage was just the guy to do it. He helped us create an Icon that would be simple and straightforward but provoke depth and meaning. See Figure 18.1.

It was extremely important to us that this Icon was done right. It had to portray our message the right way to the right people. Eric did a great job meeting our criteria, and our community loves this Icon.

Our list of Icons also includes people associated with the series. When our community sees actor Jonathan Roumie in character as Jesus, in a video or on a thumbnail, they recognize him as one of our Icons. Dallas as the face of our social media presence is also an Icon. When our viewers see the faces of these people and other actors or see the fish on the screen, they know it's us.

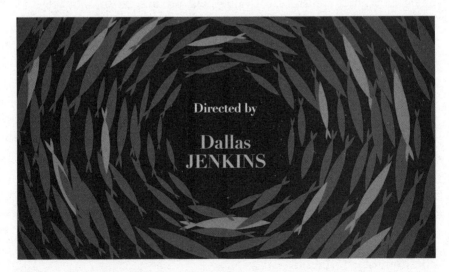

Figure 18.1 *The Chosen*

Rituals

Now that people know how to identify you, they need to know how you operate and what you do. The fourth fundamental in *Primal Branding* is Rituals. Rituals are your processes, methods, and procedures. Patrick calls Rituals meaning in motion. In colloquial terms, Ritual is "how we roll." People have seen what you're about, now they need to see how you go about it. A fun classic example of a Ritual comes from the 1960s Batman TV series. At the end of every episode, the narrator would say, "Tune in tomorrow! Same bat-time! Same bat-channel!" This Ritual was anticipated and loved by the show's fans.

When it comes to *The Chosen* series, people know what to expect from us and when. We show them the process of creation—how it all comes together. We show a lot of behind the scenes. This is an unusual approach from creators during the process of creation, but it has worked extremely well for our community. It's provided a way for us to share our successes and disappointments with our people along the way. It's become an important Ritual for our viewers.

We decided to have a programming Ritual during the Covid-19 pandemic where we offered a free episode every night for eight days. Then we would do a live stream at the end of every episode with familiar Rituals that kept engagement and retention. Dallas would welcome the viewers, introduce the episode, tease a cast member interview, show the episode, talk about interesting moments from the episode, do the cast member interview, tease the next episode, and sign off. It became a Ritual for our viewers, who continued to grow as the week went on, and we even landed on YouTube's Trending page. This was a huge deal. We had been basically nonexistent as a channel, then we gained an additional 146,886 subscribers in 14 days. Other Rituals for our followers include our crowdfunding, social presence, and advocacy. Every goal reached and milestone made becomes a Ritual we share with our community.

Sacred Words/Lexicon

The fifth fundamental in community building is how the people talk. It's called Sacred Words or Lexicon. Words identify us. Whether you speak English, Spanish, Mandarin, or Klingon, that language identifies you as a part of a culture of speakers or citizens of a specific place. The same is true of communities in sports, companies, gaming, music, and online platforms. Patrick's favorite example of Sacred Words is ordering a drink at Starbucks: iced grande skinny vanilla latte. You have to know the words so they get your order right. Even families have their own Sacred Words. If I said, "I love you six times," or, "Be of good cheer," to someone in my family, it would mean something specific to us that outsiders wouldn't understand.

Patrick explains that you belong to many communities all at the same time. You could be a musician or a banker or an artist or a poker player . . . or you could be all of those things. You have different words and associations for all of them, and you can't mix them. You don't take your rock-and-roll jargon to work at the bank. When you get a new job you have to learn all the vocabulary and jokes and learn to fit in that space.

Creed and Sacred Words can blend, so our "Get used to different" Creed also belongs in *The Chosen*'s Lexicon. People who don't belong to our community don't understand what it means. Knowing the right words proves that you belong. If you don't get it, you're an outsider. It's actually great to have outsiders—when a community's Lexicon is shared verbally or virtually or on merchandise, it allows outsiders the opportunity to ask what it means. Community members can then spread brand awareness, which brings in more people to the group.

Dallas came up with a little jingle to encourage people to contribute to or share *The Chosen*. He would sing the phrase, "www.

thechosen.tv. . ." This evolved into "www.thechosen.tv forward slash merch" or ". . . forward slash store." These jingles are a part of our Lexicon. We've even had fans send in videos of themselves singing the jingle. Another thing Dallas does sometimes at the end of videos is remind our audience of our Creation Story and our Creed by saying, "It's not your job to feed the five thousand, it's only to provide the loaves and fish."

Nonbelievers

Sixth, what would a community be without the haters. In *Primal Branding*, they are called Nonbelievers. They are the counterculture to who you are, what you believe in, and what you do. In *The Chosen*, the Nonbelievers aren't necessarily non-Christians, but they don't believe in the project or the message. Some of our biggest Nonbelievers are Christians by name but they don't agree with our message or delivery. We have had a lot of backlash over our poetic license. Some people think it should be entirely historically accurate according to their version of the Bible or their religion's teachings.

Our team has been extremely meticulous when it comes to *The Chosen*'s storyline. We want to tell the life of Jesus Christ as it is found in the gospels of the Bible, but we also refer to historical works and scholars and to leaders and scholars from many different religions. We are very careful to keep our message nondenominational. No matter how carefully we tread these waters, however, there will always be haters. And we are okay with that. When people hate, people defend. Polarization often can feed energy into a community and help it thrive even more. People become more invested in the content when they've defended it. Classic examples of community versus haters include Democrats versus Republicans, or Apple versus Android.

Leader

The last element in *Primal Branding* is having a Leader. Someone needs to be the voice of the community. Brands and channels that perform well have a defined Leader. Leaders send a rallying cry, and loyal followers will answer. I finally convinced Dallas to become the Leader or the influencer for *The Chosen*'s community even though he fought me on it for a couple of years. He conceded and went on a live stream, asking our followers to buy merchandise. He sang his little jingle, and wouldn't you know that we had a lot of money come in fast. It was a powerful affirmation to Dallas that he needed to be the Leader for our brand. He is our Leader in so many ways: he is the director, he is the writer—it just made sense that he needed to be our Leader. This was the trigger we needed to really grow and get our message out to the masses. Other examples of community Leaders include Bill Gates for Microsoft and Elon Musk for Tesla and SpaceX. They are the face of the brand, and it totally works.

Once you pull these seven fundamentals together, you are able to create a narrative that people can identify with and connect to. You can build a community around products and services, around personalities or social media influencers, among coworkers or employees, and more. If you have a belief system and you can share your beliefs in a way that builds a community, then you have the advantage over your competitors who aren't community building effectively.

Distribution

In a medium.com article Patrick wrote in 2015, he adds that there is an eighth fundamental to community building: Distribution. You have to get your message out to the people. Where are you going to do that? What elements work on what platform? Do you tell your Creation Story on Facebook? What do your Icons look like both on physical products and online? Does your community spend a lot of time on Instagram? Twitter? Learn where your people are and how to implement your branding strategies there. There are social platforms and online communities where your target audience is already congregating, so go where they are. This concept seems obvious, but I want to reinforce it because it's extremely important that you go where your audience is. If you have a gaming audience, you don't go to Pinterest, you go to Twitch. A lot of brands are still trying to fit a square peg in a round hole, thinking that if they push it hard enough or spend enough money, it will work. Don't push your content somewhere that it won't work. Go to where your people already are.

The Ninth Fundamental

We've gone over the seven (or eight) fundamentals of community building, but there is an important element we still need to address. It's important enough that I would even call it the ninth fundamental of *Primal Branding*. When you have covered all the steps and built a loyal community around your content, there should be an organic acquisition of people in that community who take it a step further. These people do more than consume your content; they become superfans. They are basically your fan club president type. They become evangelists, facilitators, and organizers for your brand. These are not the people who simply "never miss an episode"; these are the people who spend their spare time preaching what you preach and even creating their own content around your brand. These are the people who make fan art and send it to you. They build LEGO replicas of one of your Icons or scenes from an episode. They organize events around your brand. They own your vision as if it had always been theirs.

We reached out to a guy who had become a follower and advocate of *The Chosen* and asked him if he would help moderate comments for us. He ended up suggesting we create a Facebook group and auction off signed merchandise from *The Chosen* to help our crowdfunding effort. It was a great idea that generated tens of thousands of dollars for us. We would not have made that money if we hadn't integrated this fan-turned-advocate into our community building. Embrace them and reward them, because they will help you grow.

The ultimate example of superfans are well known by their nickname, "Deadheads." They are the Grateful Dead band's loyal followers who became an organic subculture in their own right. They married into communal ideals and Rituals as they followed the band on tour together, sometimes for years in a row. They even had influence over the band's song choices in real time on stage. The band allowed Deadheads to tape concerts, which was unique in the pre-smartphone era. Concert tape trading became a Deadhead Ritual that prolonged the group's allegiance and longevity even after the band's frontman Jerry Garcia died and the band dissolved. Being a Deadhead was a lifestyle, not a weekend concert-going activity. Not that Deadheadism should be your goal, but this is the kind of super loyalty you want to keep in mind as you create your own following.

Let's look at it from a story element perspective. Do you remember the pattern of narration from Chapter 13? After you tell a good story, your knock-it-out-of-the-park element is the "goosh," or the bonus. It's the pleasant surprise at the end that hooks them as a repeat viewer. If we related your community building to your pattern of narration, the superfans would be like the "goosh." They are the bonus feature that makes all the difference to your brand. You'll know your community-building campaign is a success when you see these people take their place in your community. When you convert these people, your movement continues without the Leader; they are an organism all by themselves. If you keep feeding their passion with new content, even a variety of content across all mediums, they will take your content and push it to the world. You gotta feed the fandom. Validate

them and let them feel seen and important to you. Give their fan art
a shout-out, tell them you appreciate their love and support. You have
to recognize them, or they will disconnect.

The People Are the Brand

Your community is your circle of joy. They are emotionally attached
to you or your brand, your beliefs, and your message. To be clear,
you are not manipulating anyone. It's actually counter to traditional
advertising methods that do manipulate. When you're trying to build
a loyal community, don't change who you are, just listen to your people and respond by giving them more of what they connect to.

Former clients Steven Sharp Nelson and Jon Schmidt, aka "The
Piano Guys," have an extremely loyal community. Their YouTube
channel has 1.9 billion total views, and their videos are watched more
than 3 million times every single day. Their popular videos show
them playing beautiful music in beautiful settings all over the world.
I saw them do an amazing performance for a video, and then we
decided to film a funny outro with a fight scene. The cello bow was
used as a sword, and the piano used as a face-pounding surface. It
was really great content. However, when Steven and Jon talked about
it, they decided not to include the outro. I pushed to keep it because
I knew it was viral-type content, but they pushed back. They knew
their community, and they knew this was the wrong content for their
community, even though it was great content. So they did a regular
"boring" ending. Kudos to Steven and Jon for protecting their message and their fans. This taught me more about staying on brand and
not diverting from your message, even when you have content that
could be a big hit.

Communication in consumerism used to come from the top
down: the manufacturer would tell the consumer about the product.
But now it's flipped. The consumers are on top and saying "Hey, I

have the money; come and find me," and the brand has to respond in order to get the business. Patrick says, "The biggest shift today is that brands used to be about products and services—but today 'Brand' is about the people who buy those products and services." It's all about the community.

Patrick spoke at an event once and there was a guy in the audience who really took his message to heart. He returned to his office and figured out how the seven fundamentals of *Primal Branding* related to his business. He transformed his company by building a community. The company grew exponentially and sold five years later for $12 million. The buyer then turned around and sold the company for $150 million nine months later. They attributed the value of the company to its community, stating that it couldn't be replicated.

Primal Branding's message is clear: believe in what you're doing. Know where you're from and what you're about. Figure out a way to identify and declare yourself so others can find you and identify with you. Create a Ritual for others and a vocabulary or language for your own community. Make sure people know what you are not about. And lead the way. You don't have to have a community as big as Apple Inc. to change the world, you just have to be unified. If you have the right components for a community, you'll succeed, and you can really make a difference no matter how big or small your channel is. Some channels I just don't get, so I'm not a part of their community, but there are a lot of people out there who do get it and do want to be a part of it. The same is true for groups I do care about that others don't. We all have a place; we all belong somewhere.

I encourage all of you to analyze your community. Have you developed loyal followers? Are you utilizing the fundamentals of building a community? You might think you don't have the right genre or the right content to gain loyal followers, but remember that even Squatty Potty found a unique audience who would share their message. If the Bronies can have their own community, so can you.

Action Exercise

Task 1: Using your viewer persona, develop a plan to build a community with your most loyal followers. Ask yourself:

How will you engage with your audience monthly?

What action steps will your community take each month?

Task 2: Recognize and reach out to some of your most loyal followers and ask them to be your channel moderators.

Get the companion workbook and find more resources at www.ytformulabook.com.

19 Optimizing, Launching, and Promoting Your Video

Most content creators upload a video and just let it do its thing, but the upload is not the end. If you want to really succeed on YouTube, you have to actively watch and respond to the data coming in after upload. Your #1 goal is to get the right type of viewer coming in quickly so the algorithm can recognize them and know who to keep pushing it to. You should treat the upload as a part of a sequence when you're launching a new video. Let's walk through step-by-step on uploading a video to YouTube. Then we can get to the fun part of optimizing, launching, and promoting your content.

How to Upload a Video

In your YouTube Creator Studio, click on the icon in the upper right corner that gives you the option to upload a video. Then you can upload your video in one of two ways: either drag and drop, or click on Select a file. You can actually upload as many as 15 videos at once here, which can be a huge time saver. How long it takes to upload obviously depends on your file size and your Internet speed. Once it's done with the processing time, you'll need to change the default title to the custom one you've carefully crafted from the exercise in Chapter 15. Then you need to fill out the description, which is also important to include words that deliver on the promise of the title and video. Next up, change the auto-generated thumbnail to the custom one you've spent a lot of time creating (also from Chapter 15).

After you've plugged in your amazingly clickable thumbnail, you can choose which playlist or playlists to add your video to. Then you'll see a fairly new upload feature: choosing your audience. YouTube added this step because of their issues with the Children's Online Privacy and Protection Act (COPPA). They just need to know whether your content is geared toward kids. In this feature you can also designate if your content is for a mature audience. It is very important that you fill out this part correctly so you don't get demonetized.

At this point on the screen, a lot of creators skip the next part, which is "More Options." Go ahead and click on that so you can designate if your content contains integrated brand deals, sponsorships, or other paid promotions. Here, you also add tags. Include tags with keywords from your video's title and description, for starters. Then you can choose your language and closed captioning (CC) preferences. The next option I usually skip, because it's asking for the recording date and location of my video, which doesn't really matter; it doesn't affect the algorithm or your video's exposure. In the license and distribution section, you can choose the Standard YouTube License, which protects your content from copyright infringement, or you can choose Creative Commons, which gives anyone free reign to use your content without legal repercussions.

It's important to make sure the next two boxes are checked so you can embed the video on other websites, but more importantly, so your subscribers will be notified that you have a new video. Now you're ready to choose your category and your comments preferences. If you want to hold potentially inappropriate comments for review or turn off comments altogether, check that box here.

You're all done with the Details tab at this point, and you are ready to go to the Monetization tab. If you have monetization available, you can choose exactly what type of ads you want to run with your content, whether it's overlays, display, or skippable/nonskippable ads. You also can choose if you want ads to run before or after the video, or during the video if it's longer than 10 minutes.

If you are a YouTube Partner and can run ads with your content, the next part is important. It helps you narrow down your ad suitability. Again, fill this out correctly so you don't get demonetized. You can specify exactly what is in your content as far as adult content, violence, harmful or dangerous acts, drug-related content, hateful content, firearms-related content, or sensitive issues. There is an option here to click None of the above, but don't click it if there is anything at all questionable in your content because the algorithm will find it, and you'll likely get a Community Guidelines strike, which you definitely do not want. Again, don't risk getting demonetized.

We're almost there! You can now specify what video elements you want added, like endscreens and cards. Endscreens run at the end of the video, while cards can be added anywhere you choose. Once you're done adding what you want where you want it, click on the Return to YouTube Studio button.

You've made it to the last tab: Visibility. Most of the time, you want your video to be public, but there are times when you might choose one of the other options to make it private, unlisted, or for members only. You'll also see the video link on this screen. You need to click on that link and go to the settings on the video to check that it's been fully processed and is ready in high resolution. After checking it, you can schedule your video to be released on a specific day and time. I almost always schedule videos. It gives me the freedom to work on uploads when I have the time to, so I'm not a slave to the clock.

One awesome service I recommend to every creator is called TubeBuddy. TubeBuddy is an extension that gives so many helpful tools to optimize your videos and save you a ton of time. When you sign up and install it, you'll see all of the amazing things it can do to help you streamline the upload process and optimize your video.

How to Optimize a Video

Optimization is all about humans and all about the click. I could say a hundred times that the algorithm's job is to predict what the viewers will watch. It follows people around seeing what they click on and what they don't click on. When you have a higher click-through rate (CTR), that tells YouTube that people are actually interested in your content. Get the people interested in your content!

Ninety percent of all top performing videos on YouTube are using custom thumbnails. This is why creating a good thumbnail should be your first consideration. Put in the energy and time to make it great. This takes more time than you realize . . . then probably more.

Use colors to make your thumbnails really pop. Use words when necessary. Use emotion, because it creates that curiosity you want in a potential viewer. Always be honest. Your thumbnail and title need to actually reflect what the video is about, or you'll probably have a low average view duration (AVD). Use catchy, clickable titles that hook humans (not robots). A good rule of thumb for titles is to be catchy but concise. Put your main or primary keyword as close to the front of the title as possible—this helps you get ranked and suggested. Of course, your title has to be relevant. Sensational, irrelevant titles create low AVD. And don't forget about your buckets! Use words in your title that connect a video to others in the same bucket, or category.

The right keywords in your description will help you get more views and visibility in search. "Above the fold" is the most important part of the description because it's what people see first and what the algorithm sees first. "Below the fold" means under the words "show more." People have to click here to see below the fold. You only have 200 words above the fold, but you can say a lot more below the fold. In the description, tell people what to expect, but always write like a human. The algorithm likes humans, remember. Below the fold is a great place for links to other videos, links to your socials, relevant tags, and timestamps. A word of warning here, do not reuse the exact same description over and over; the algorithm doesn't like it.

Some creators don't optimize with cards and endscreens, which I think is crazy. Why wouldn't you want to give your viewer more ways to watch more of your content? However, don't make the mistake of putting a card in the first third of the video because you don't want the viewer bouncing too soon and ruining your AVD. In your cards and endscreens, choose a video from your library that the viewer is more likely to be interested in. I call this the target video. Think about your target video in preproduction along with your title and thumbnail, because connecting your data and your buckets is your ultimate goal. I also use a feature on endscreens called "best for viewer" that lets the algorithm choose a video from my library that it thinks that particular viewer is likely to watch. Use this to supplement, not replace, your target video suggestion. In addition, take the last 8 to 10 seconds to use a verbal call to action asking them to watch the target video. Even if the percentage of people who click on your suggested cards and endscreens is low, it's still activity that the AI is watching and will reward. And guess what happens when they do click on your suggestion? The algorithm recognizes the connection and puts it in the Suggested feed.

Write relevant tags. Tags are less important than your title, thumbnail, and description, but they still help, especially if some of those words are often misspelled by people. For tags, use your primary and secondary keywords and other relevant words. Stay under 300 characters in your tags. TubeBuddy is a great tool to show you relevant tags based on your current video's content.

Power tip #1: Use closed-captioning (CC). But not YouTube's auto CC, because it can hurt your ranking and CPMs; it's just not very good. There are services you can pay to do your CC very inexpensively. CC helps validate the work being done by the AI with video intelligence and natural language. CC will increase your Watch time, too.

Power tip #2: Use default profiles on TubeBuddy. This has saved me hours and hours of time over the years. It allows you to save your most used tags and descriptions, which you can go in and tweak depending on the current video, but you won't have to start from square one.

How to Launch Content

Remember Dan Markham's release strategy for the launch of his soccer video in Chapter 17? That was a great use of collaboration, trending, tent poling, and even timing his launch down to the exact hour to maximize his audience and success. That's great for launching a video, but you can take this further when you launch a channel.

I have used MrBeast over and over again as an example of what to do. He just knows the Formula and does it right. MrBeast's main channel is doing amazing, but he was ready to start a gaming channel. He wanted to create his new content around *Minecraft* and gaming, because that's where he got started and spent the most time. We started to do our recon and research with comparative analysis on similar gaming channels. We were extremely sensitive to the YouTube Formula and finding the right viewer. He was very careful not to push the audience from his main channel to his new gaming channel, because they were different avatars. If you get the wrong viewer watching your content, it messes with the algorithm.

We started creating and testing content months out from channel launch, and our ideal avatar ended up being different from what we had thought it would be. We scrapped our previous strategy and content based on the data we had coming in, and we reset, creating content geared toward the ideal avatar according to the data. A month before launch, we had developed who the ideal avatar would be, and we created relevant content with calculated title/thumbnail combinations. Our main focus was to get the right click. Our A/B split testing helped us identify what could work. We were able to use people who had a lot of success in the niche because of our YouTube connections. MrBeast's regular pacing and edits wouldn't work on this gaming channel, so it was interesting to watch him pivot and do something completely different for this new channel.

We decided to release new videos three times a week, which is pretty aggressive, but not for a gaming channel. We knew we had enough content to start pushing, so we went for it. We needed to promote the content, but we waited until the third video to start promoting. All we did was ask some big gaming creators to share our videos in their Community tab. These were creators we wanted to collaborate with. When we did this, our views went from 200,000 daily views to 1.1 million views in one day. We had leveraged the algorithm and used the YouTube Formula, and it worked. The next day, we got 1.6 million views, and the next day, 1.7 million. We had triggered the algorithm to put our video in front of the right viewers. Revenue started coming in, too.

This was for a channel launch, yes, but it works for a video launch, too. Give the algorithm exactly what it's looking for. Don't confuse it by pushing to the wrong viewers in the wrong places including your social media friends and followers. Of the 10.6 million unique viewers, they watched four videos apiece, and 1 out of

10 viewers became a subscriber. These numbers are crazy good. You need to understand:

- Your audience
- What makes good content
- Traffic and momentum to drive good CTR
- How to grab their attention and keep them engaged
- How to use the buckets system to recommend similar content so it will get suggested

When you understand these things, then you understand the Formula to leverage the power of YouTube.

How to Promote a Video

I have been on YouTube since 2005 and there's one question about growth that I hear year after year: Why can't I just run ads to grow faster? The answer is a complicated one that I hope to simplify in this chapter. If you want to expand your reach, do a product launch, or sell merchandise, running ads can be really effective. But there is so much more to successful campaigns than dumping money into an ad.

I have seen a lot of big brands throw money at a problem and hope for a magical solution. Their views can go up for a minute, but that video doesn't necessarily convert. This can be a vanity metric because the ad does produce some results—it's like a virtual pat on the back. When I do work with clients through my agency, I do split test after split test running ads against organic content. For any job that I do, I want to create an ad that will convert 100% of the time and not be a Band-Aid for a numbers problem. I do not advocate putting a Google paid strategy solely around one piece of content. It is a tool that can amplify what you do, but you have to do it right.

Market organically and supplement with paid strategies to have a higher probability of success.

Promote Organically First

When I helped produce Squatty Potty's viral ad "This Unicorn Changed the Way I Poop," we combined both organic reach and paid strategies. We planned and tested everything meticulously. Our final cut performed at a higher rate after launch because we had run several tests to know what content performed with what audience. If you want to grow a channel, you have to figure out who will click on the video.

To figure out an audience, I like to look where they congregate both when they're online and when they're not. I also want to know what content and products they connect with. This helps me know what will be relevant to get our content out to more people. I also want to know what the client's endgame is: when a client has a goal to make a sale, I use paid strategies differently than when their goal is to grow an audience. I would rather spend money on 100 relevant channels to push organically rather than throw money around randomly. This is how you get dedicated people to engage with your content.

Paid Strategies to Grow an Audience

If the client's goal is to attract more viewers, I go find a similar creator who can collaborate with my client and organically push the right viewers to our new content. In YouTube analytics, you now can see your crossover viewership listed in "Other videos your audience watched." This makes it easy to see who you can reach out to to develop a collaboration relationship with. See Figure 19.1.

I also look at the right influencers' older content that is relevant to my client's new content, and we pay them to add endscreen

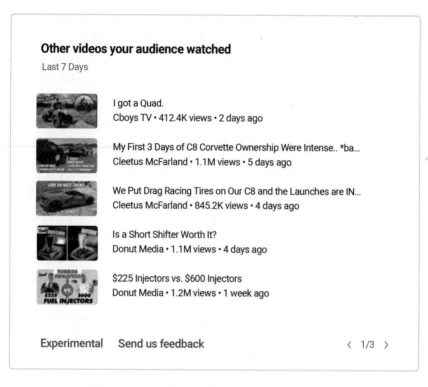

Figure 19.1 Other videos your audience watched

elements, links, and shout-outs that refer their audience to our content. When a loyal viewer of a certain influencer watches that person's content to the end, and that influencer pushes your content or product, you are much more likely to get what you need from that viewer because they came organically through YouTube's ecosystem.

In addition, use your Community post or Story to push more views to your new content. You need to have a blended strategy like this to push your content to a diversity of traffic. When the algorithm sees that viewership is finding your content from multiple sources, it will recommend your content more.

Let me give you the perfect example of a great blended strategy. When BYUtv's channel *Studio C* was first starting out, I was helping them with audience development, doing recon and research and using every tool possible to try to figure out who their ideal viewer would be. I reached out to identical-twin sister creators Brooklyn and Bailey to collaborate with *Studio C* because I noticed we had some crossover data. The girls were actually big fans of *Studio C* and were open to a collaboration. We offered to do a "one-for-one" collaboration, meaning we would make two videos: one on their channel that we would produce instead of them, and one on ours. At the time, Brooklyn and Bailey had 300,000 subs and *Studio C* had 20,000. By the time we were ready to launch our campaign, Brooklyn and Bailey were already nearing a million subs.

Brooklyn and Bailey's video, "The Other Parent Trap," parodies the movie *The Parent Trap*, with the twins discovering their sisterhood at a summer camp and switching places. Brooklyn arrives at her dad's house and is welcomed with open arms because Dad is none the wiser, while Bailey gets arrested immediately because Mom knows she's not Brooklyn. *Studio C*'s video, "Prom Dress Gone Wrong," features Brooklyn and Bailey alongside the show's actresses, all wearing ridiculous prom dresses, to their dates' collective dismay. This video currently has 11 million–plus views.

In addition to the two videos being released, I used a really interesting remarketing strategy: I took a clip of the girls talking about how much they loved *Studio C* and used it in a paid ad campaign. In 12 days, *Studio C* got 162,000 subscribers that came from Brooklyn and Bailey's recommendation. We also had campaigns to benefit the twins. It was a great win-win setup.

We had a small budget to do another ad remarketing strategy, which we pushed strategically to the right viewers. Another 62,000 subscribers came from this push. So, we did it again. Why not? It was working! We waited another three or four months to build data, looked at the data, and targeted another specific audience with a paid ad strategy, which brought in another 12,000 subscribers.

Fast-forward a year. We decided to do another one-for-one video campaign that was similar to the first collaboration. We created two new videos and followed the same strategies. This time, the majority of our budget went toward content creation. On Brooklyn and Bailey's channel we did a video called, "High School Boyfriend Drama," with one of the *Studio C* actors playing the boyfriend who got dumped by one girl and immediately found another girlfriend who looked just like the first. On *Studio C*'s channel we made a video called, "Dance Battle Boys vs Girls," and included Brooklyn and Bailey as part of a girl group who has an awkward dance-off with a group of boys (all *Studio C* actors, of course). We released the videos at the same time and watched as our views grew into the millions, yet again.

Brooklyn and Bailey pushed traffic to us organically by doing shout-outs on their social media pages. Our ads were targeted specifically to the viewers who were subscribed to Brooklyn and Bailey but not to *Studio C*. We got well over 80,000 subscribers this time around. The goal to grow an audience with a blended paid strategy had been achieved.

Paid Strategies to Sell

When the goal is to sell products, most creators actually get it wrong. They don't know how to push the product launch in an organic way while also pushing it with a promotion strategy. Don't be so arrogant in thinking that your viewers will watch to the end of your video where the product push will be. You need to give them several ways and places to see the product. Spend your budget the right way, which is developing and understanding your content and making it perfect for your ideal avatar who becomes your ideal customer. You want it to feel natural and organic even when you're paying for it. It's the best way to get people to do what you want them to do, which is usually to subscribe, watch more, or buy. In this case, buy.

One of my favorite examples of a successful paid strategy to sell comes, again, from *Studio C*. In 2014, they had a soccer sketch with a character named Scott Sterling who took a soccer ball to the face over and over in a shootout. The initial video, "Top Soccer Shootout Ever with Scott Sterling," had gone viral. I tried to encourage them from the beginning to sell merchandise, but they didn't do it. So they wanted to do a follow-up video with Scott Sterling in the hospital, and this time, their goal was to sell T-shirts. The video, "Scott Sterling Breaking News Update," was released almost five months after the original.

In this video, a nosy news reporter storms into Scott Sterling's hospital room and begins asking him questions about his plans to return to soccer in the future. Sterling's trainer intercepts the pesky reporter and defends him. The video is just over two minutes long and includes *Studio C*'s regular comedic pace, which makes it feel natural to their regular, ideal viewer. But 47 seconds into it, you'll see the first call to action box "Click to Buy" for the T-shirt. The trainer then pulls the call to action into the script, keeping the viewer engaged by adding the comedic element to the CTA. They finish out the sketch, and at the end, the CTA box pops on screen again. The trainer adds sarcastic comments to the merchandise push, telling viewers not to buy it for their friends and family, mentioning the quality of the fabric and the reasonable price. We spent just a few thousand dollars on this promotion, and we made millions in T-shirt sales.

As you can see, promoting with a blended strategy of organic reach and paid push produces the best results. This is true whether your goal is to grow an audience or to sell something. You always want to do your homework on the people you're trying to reach. Then you can find other creators or businesses with crossover data who can help you reach those people organically. I've done it time and time again with clients in all different genres. Do it right, and you'll get the conversion you're looking for.

Action Exercise

Task 1: In your next video, use the strategies from this chapter to optimize, launch, and promote your video.

Task 2: Look into your analytics to see what other videos your audience watched. Make a list of 5–10 creators you would like to do a collaboration with.

Get the companion workbook and find more resources at www.ytformulabook.com.

20 Tweaking Your Content

We've talked about how to create engaging content, good thumbnails and titles, and how to get your video on YouTube. In Chapter 19, we went over practices to optimize and promote your video. Now you're ready to learn what to do after your video hits the shelf. Do you know what to do if your video is underperforming? Do you even know how to tell if it's underperforming besides views? What if it's taking off? How do you respond?

I've said it a thousand times: you have to analyze and adjust; analyze and adjust. In theory, this means you have to improve your content. In practice, it means you look at the data in your YouTube analytics and see what is going on and what you actually can do about it. You have to assess your progress. This is true for every single YouTube channel, no matter how big or small or old or new. This is where the rubber meets the road with everything you've learned in this book.

Certain content is good at gaining new subscribers, other content is good at retaining subscribers, and some content is good for both subscribers and nonsubscribers. Figure out which kind of content you're dealing with and who it's geared toward to know which tweak to make when it underperforms. For example, I was at a private retreat in Las Vegas with MrBeast and other data-driven creators.

We were talking about MrBeast's content and his viewers. I asked him which videos he enjoyed making the most, and he said he liked the videos where he helped people in need. MrBeast had a lot of videos giving away money and big things, but often, he was giving it to his friends.

I looked at MrBeast's videos and how they performed on average. I noticed that the videos of him giving away things to people in need rather than to his friends actually performed better. He had gotten 4 to 10 times more new subscribers from those videos. He needed to make more videos like that, which was awesome, because that was his favorite type of video to create anyway. This was a small tweak that brought a huge return in subscribers and viewership.

Real-Time Tweaks versus Long-Term Tweaks

You need to analyze and adjust in two ways: in real time, and in strategizing for future content. In order to do this, you have to know how your content is performing in relation to how it performs on average. How do you analyze if your video is performing well? When you do know how to analyze, how do you adjust to fix it? I'll show you how to find your average numbers, also called baselines, so you know how your new video performs in relation to how your content performs on average. Then I'll show you what to do to adjust for outliers—videos that land either above or below average.

As you learn to track your baselines, you'll learn to recognize patterns. Patterns tell you what is working and what you need to change so your content will perform better. We're not talking about huge changes and adjustments. They are small changes, so we call them "tweaks." Small tweaks can translate to big peaks in your metrics graphs. It's like the butterfly effect: it's the idea that a butterfly flapping its wings on one side of the world can significantly change the weather on the other side of the world. Your small tweaks can significantly change the trajectory of your video. But you have to know your baselines in order to know what to tweak.

How to Determine Your Baselines

(Note: If you have a new channel and don't have the data you need yet, go back to recon and research and look at another channel that is doing well in your niche and analyze what's working so you know what you should be doing.)

Most creators try to determine their baselines from a whole channel perspective, but this is a huge mistake. A whole channel approach doesn't give you accurate averages to be able to leverage who is watching from where. To illustrate this point, let's say you're going to school and you have five classes. You have really good grades in four classes, but you're failing one class. If you only look at the average score of the five classes combined, it looks like you're doing fine. This overall score can't show you that what is bringing the average down a little is actually a failing grade. But if you break down the grades by class, you'll see that the one class needs some serious attention, stat. This is why you need to go from a macro view to a micro view when determining your baselines and knowing which content to tweak. In order to do this, you have to break it down to the traffic sources level.

YouTube doesn't provide baselines for you based on traffic source; you have to go and find that data manually. When you upload a video, you need to track the most important metrics by traffic source. The most important metrics are:

- Click-through rate (CTR)
- Impressions
- Average view duration (AVD)
- Average percent viewed (APV)

- Watch time and views

- Average views per viewer (AVPV)

I like to have data for at least 10 videos so I have enough data to provide a good baseline. To establish your baselines, select a traffic source. Then select 10 videos in that traffic source. So if you did this with Suggested traffic and CTR first, you would get your 10 videos, add those numbers up, and divide by the number of videos. This is your Suggested traffic CTR baseline. (Note: If you've established your video buckets, group your videos by bucket and then find your average.) Repeat this process for the other traffic sources for CTR. Then move on to the next metric and repeat the process of finding its baseline in each traffic source, as you did for CTR.

CTR and Impressions baselines are the most important ones to establish, especially for real-time observation. However, you also need to find your baselines for AVD and APV for long-term observation. This is important because even if you have a great CTR, if it doesn't produce good AVD and APV numbers, the AI won't suggest your content to a more general audience. YouTube wants to push your content to the widest possible audience so your content can get more impressions and more views. The more impressions you get, the lower your CTR will be because it's going out to more viewers.

What to Do with Your Real-Time Baselines

Everything you do in YouTube analytics stems from knowing your metrics baselines. Don't fall into the comparison trap, always measuring your data against other creators. Other creators' numbers can't match up to yours because there are so many factors determining those numbers. Focus on your own numbers only. There are dozens of ways to look at metrics in your analytics, but stick to the most important ones we listed above. Do the following with your CTR and Impressions data:

1. Look at your CTR as soon as YouTube gives you that data, generally two to three hours after uploading a new video. This is your real-time data.

2. If your CTR is below the baseline you've established, your next step is to look at your Impressions data before you change anything. This is super important because if your impressions have gone up, it means the AI is pushing your video to a broader audience. More eyes get the chance to see your thumbnail and title, which is a good thing, so even if this makes your CTR number lower, your video is actually getting seen by more viewers! This is a hard concept for a lot of creators to grasp. Switch your mindset from "Low CTR = Bad," to "Low CTR = Could be good, let's look a little harder."

3. If your CTR is low but your Impressions are high, don't change anything yet. Let your content simmer, and see what it does. But if your CTR is low and your Impressions are average or below, it's time to adjust in real time. You need to change a thumbnail or title now. I recommend switching out your thumbnail first because it's what grabs viewers' attention before they look at the title.

Most creators go wrong here and look at things that don't even matter, or they don't look by traffic source. Don't overwhelm and confuse yourself by looking at too much. You'll learn how to look at these baselines and understand what you need to do to adjust and get results. Just remember that in real time, the most important baseline is your CTR. Everything in the algorithm machine stems from here. If you haven't been making this metric a priority, do it now.

Have Backup Plans Ready

Here's where I differ from most creators: I have a plan A, plan B, plan C, and plan D all ready to go when I upload a video. Some people think this is overkill. Maybe it is. The point is that you have to have thumbnail options ready to switch out as fast as you can if needed. When you can course-correct quicker, the algorithm will see that you've made a quick change and respond accordingly. I worked with a client who had a great thumbnail and a great video. Before he uploaded it to his channel, he asked our Discord group for feedback and got some great advice. But when he uploaded the video, it was falling under his baseline, so we switched out the thumbnail, not once or twice, but three times. Finally, the video took off, and he got 50 million views. *Fifty million.*

Do you think we regretted having four thumbnails ready in advance? The same thing happened with another client: he uploaded a video, and it just wasn't performing well. He waited two days before changing the thumbnail, and the video did pick up, but if he had changed out the thumbnail two hours after upload instead of two days, it would have done so much better.

Let me expound. When you upload a new video to YouTube, that video's clock starts ticking. It's "aging." Its freshness factor is on a steady decline from the outset. Kind of like that saying about buying a new vehicle: it starts dropping in value the second you drive the car off the lot. This is why it's imperative that you respond quickly to the data that comes in about a video's performance. You won't be able to capitalize on the Browse features of Homepage, Trending, and Subscription feed if you don't act quickly. These are your "act fast" metrics.

What to Do with Your Long-Term Baselines

When you start watching your AVD and APV baselines, which are usually available 48 hours after upload, notice if they are below or above average. Then go to where you lose viewers in the video or where there is more engagement in the video and try to find a pattern for why that is happening at that point in your video (learn more about AVD and APV in Chapter 16). I've done this thousands of times, and there is always a pattern to discover. When you can identify patterns, you'll know what to adjust for future content creation.

The Suggested feed requires slow and steady consistency and small tweaks. In your analytics, click on Reach, and look at your highest traffic source for the last 7 days, 90 days, and year. This is where a lot of YouTube channels go wrong; they only look at the most current video and what's happening with it right now. But when you look at how your content performs over time, you'll see different patterns.

Even if your highest percentage of traffic over time shows Browse and Subscription, guess what? You can make small tweaks that, over time, help your channel get higher and higher in Suggested traffic. You don't have to be stuck as a Browse traffic–only channel, and you don't want to be. Your Browse feature traffic (Homepage, Trending, and Subscription feed), are your most valuable group of viewers, and you need to make decisions to cater to them. They are your loyal followers. But if you're not also working to improve your Suggested percentage, you're not doing it right.

Split Testing

I have used a tool called TubeBuddy for years, and I recommend 1,000% you use it with your channel, too. You can download TubeBuddy at www.tubebuddy.com/go. It will save you so much time and energy, not to mention it will help you figure some things out with your content. An invaluable feature that I love to use is the A/B split testing. To run a split test, click on a video in your library and add a secondary thumbnail with a subtle variation, like changing a background color. It doesn't work as well if you use a completely different thumbnail variation. After the test has run its course, look at which thumbnail performed better, and more specifically, where the traffic was coming from. Be careful not to take a quick glance and make a rash decision; the numbers can be deceiving if you don't look at them right. Use your brain and dig a little deeper. See which traffic source is improving, and if that's your goal for this specific video.

Small Tweaks

We spent a whole chapter on how you get people to click or tap on your video. You have to have good titles and thumbnails. Getting people to click triggers data to start feeding your metrics. Then you'll be able to see how this video stacks up against your average performances so you'll know if you need to change something. Changing something usually means swapping out a thumbnail image first to see if that does the trick. If you watch your baselines after the swap and it doesn't fix it, then you can tweak your title and watch your baselines again. If this still isn't working, you can go back and swap out yet another thumbnail. Don't be afraid to do this. I've seen it happen time and time again where the second or third (or sometimes fourth!) thumbnail finally clicks and the numbers start to climb.

One of my students has a channel that is really taking off. It's called *Matt's Off Road Recovery*. Matt documents his towing rescues of people who get stranded in his geographical area. Matt came to me and my team to get help with his channel, so we looked at his metrics and helped him determine his baselines by traffic source.

The first thing we did was focus on his thumbnails. We found the baselines for all the thumbnails that had the highest CTR, and we noticed a pattern. The thumbnails that had Matt's face with an expression on the thumbnail performed three times higher. We did a few tests and determined that our hypothesis about his face with expressions was true.

Next, we needed to fix his AVD/APV. Look at Figures 20.1, 20.2, and 20.3 for Matt's channel. In Figure 20.1, the video's AVD had a big drop off. We took all of Matt's videos that underperformed and looked for patterns and hypothesized why viewers were disengaged. Then we looked closely at the group of videos that were his best performers. We noticed a few patterns in his videos that had a higher AVD and APV. We were able to test these hypotheses in a few upcoming videos, implementing strategies we thought would work. Then we looked at where viewers engaged the most.

After we looked at Matt's baselines, we were able to make small tweaks to the content over time. Look at Figure 20.2. Our tweaks brought Matt's AVD/APV back up to the average line. Notice that his APV went from 52.5% to 71.8%.

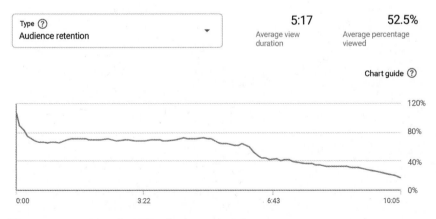

Figure 20.1 Matt's AVD drop

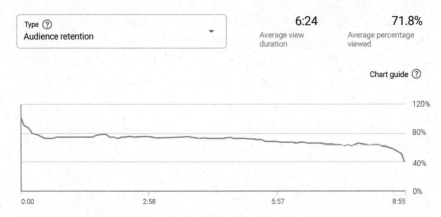

Figure 20.2 Matt's AVD recovery

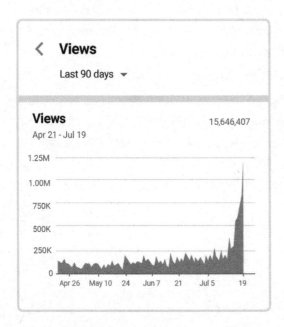

Figure 20.3 Matt's video views

Matt's video views skyrocketed. Figure 20.3 shows what our small tweaks did for his channel.

Remember that it's all about the viewer—from seeing your thumbnail and title and wanting to click (impressions, CTR), to wanting to finish your video (AVD/APV, Watch time) and watch more (Watch time, AVPV). It's about always trying to learn and improve by analyzing and adjusting.

Just Do It

The best way for you to figure out how all of this relates to your content is to dig in. Go to your analytics and learn how to read your most important metrics. Have variations of thumbnails ready to switch out quickly when one isn't working. Know where your traffic is coming from and what your baselines are. It's nearly impossible to fix a video that's not performing if you can't read your real-time data. The more you practice, the better you'll get at seeing patterns and knowing when to act fast and when to let your content simmer. You'll be an analytics pro before you know it, and your content will improve leaps and bounds as a result.

You'll figure it out. It will take time and work, but I promise it's worth it. What do you have to lose? Some people catch on to patterns and data analysis quicker than others, but no matter how long it takes to learn it, the end result is the same. You'll learn how to analyze your channel data so you can make the adjustments necessary to succeed.

Action Exercise

Task 1: Find your baselines for videos in one of your buckets by following the steps outlined in this chapter.

I go in depth on baselines analytics in the companion course for this book. Get the companion workbook and find more resources at www.ytformulabook.com.

21 Try, Fail, Analyze, Adjust: Your YouTube Success

Let me leave you with a story of an amazing YouTube family. Jase and Rachel Bennett's YouTube channel is *The Ohana Adventure*, or *TOA*, as they call it. It's a family vlog channel that provides a comfortable lifestyle for the Bennett family, but it wasn't always that way. In 2013, Jase had gotten his longboard brand, jaseboards, into Costco and other big retail spaces when an underwater skateboarding video went viral on YouTube. Jase saw an opportunity on YouTube for his wife, Rachel, who had been a regular blogger for many years. Rachel liked the idea of starting a family vlog, so *TOA* was born.

Rachel began vlogging without much direction. In the beginning, she didn't even watch other YouTube channels to see what other vlogging families were doing. But she did do something powerful: she found her Why. She wanted to be one of those families who could inspire others to nurture healthy family relationships by having fun together. It took nine months to get any views, and when the channel hit a hundred subscribers, it felt amazing (as good as a million subscribers would feel down the road).

One video in particular boosted *TOA*'s YouTube success. Rachel did a "How to Shave Your Legs" video with her preteen daughter.

This triggered a nice run with monetization that made Jase decide to quit his job and go full time on YouTube with his family. The Bennetts filmed everything in the regular life of a family, including the "ah-oh" moments, like a kid bonking his head or getting another injury. Then along came the Children's Online Privacy Protection Act (COPPA) and a super strict response from YouTube: all "questionable" kids' content got demonetized, including *The Ohana Adventure* because of their "ah-oh" content. This was a hard blow to the Bennett family, whose livelihood depended on their YouTube income.

I had a conversation with Jase and Rachel at this time. We talked about the importance of creating data relationships among their videos and getting their content in the Suggested feed. We talked about collaborating with other family vloggers. They made a new plan based on these ideas, and they committed to it. Rachel was determined to get her channel back on track, so she gave herself a deadline to reestablish monetization, and she got to work making better content.

Jase and Rachel surrounded themselves with like-minded people and did some great collaborations. They even shared audiences, topics, and upload schedules with similar family vloggers. They made sure their audience had good crossover data with the channels they chose to collaborate with so they would get the right viewers watching their content. Every Monday night for two years they did a live stream and played family games. They knew consistency was important, so they never missed a week. This was a new strategy that gave their audience a chance to connect deeper. Viewers had great comments that gave them ideas for more videos, and they listened.

Audience engagement was huge. When they listened to suggestions and gave shout-outs, it pulled their ideal viewers into their content even more. Jase started doing a silly rap at the end of every live stream where he would plug viewers' usernames into the rap. This gave him the idea to do a video with his daughter that parodied

Taylor Swift's song, "Look What You Made Me Do." The parody became their most viewed video with more than 100 million views. Because of this video's huge success, they made more videos with data relationships to connect to that content bucket and to their overall content. They made videos of the parody's bloopers, behind the scenes, and choreography. This strategy to build more content around what was working gave them $20,000 in AdSense in the first week.

Every YouTube channel makes mistakes, and *TOA* was no exception. Jase and Rachel got caught up in the comparison trap, thinking if they started to copy what other successful creators were doing, it would work for them, too. It didn't. Jase got stuck in an algorithm mindset, creating content for the machine instead of for their viewers. They had stopped speaking to their own audience, and their channel went from getting millions of views a day to getting 500,000 at best. Their channel was dying, losing momentum and interest of the viewers.

So they regrouped. They went back to their ideal viewer, which was actually an older kid demographic, and slowly started to get their metrics back up. Now they are careful to gear new content to the right audience. The kids have separate channels from the family channel, and each appeals to different audiences based on their age and topics. Because they have grown up on YouTube, the kids know their audience well and do a good job creating content for their ideal viewer.

Jase and Rachel chose YouTube as their family's career and main source of income, but it's important to them that they keep it fun for the kids. They live and create by the mantra "Inspire, not require." They've tried, failed, analyzed, and adjusted to get to where they are now. Their collective channels have millions of subscribers—people who are inspired to have good old-fashioned fun and healthy familial relationships. The best part is they went back to their original Why from when they started on YouTube.

When I asked Rachel what advice she would give to herself if she could go back to when she started her YouTube channel, she said she would tell herself to figure out what content performed best so she could do more like it. She would be more consistent. She would have good metadata and create viewing data relationships among her content. But maybe most important of all, she would tell herself not to let the blunders get you down. Just move on, and make better content.

YouTube's own history follows the formula to try, fail, analyze, and adjust with data-driven decisions. Jawed Karim, Chad Hurley, and Steven Chen started YouTube with the intention of making it a dating website. It didn't work, so they looked at what was working, and changed their strategy to match the data. You have to follow their example, and *TOA*'s, if you really want to succeed on YouTube. You now have the tools you need to be smart in your preparation, creation, and follow-up, meaning you know what it takes to try, fail, analyze, and adjust based on data. Utilizing the data to your advantage takes the guesswork out of your YouTube channel and not only helps you succeed, it feels incredibly empowering.

You've done some great work in this book. You learned how the AI has changed over time and how each traffic source now has its own algorithm to cater to. This was a very important thing to learn, because it helped you take control of your own presence in the ecosystem rather than blaming YouTube for your failures. Now you know that there are opportunities aplenty—a whole world of views to capture and money to make. There are countless ways to interact and succeed through community, collaborations, brand deals, and more.

As you move forward as a creator, align your priorities and practices with what YouTube wants. Put more importance on getting viewers to click or tap on your video, because if they don't click, they won't watch. And when you get them to watch one, then get them to watch another one. Think of the viewer as a person, not a number, so you can really get to know who they are to anticipate what they might want to watch. It's all about the viewer! And remember that YouTube success is not a lottery. Don't think that Jase and Rachel Bennett were some of the "lucky ones" realize that anyone can do this—including you—with the right mindset and tools.

I hope I have opened your eyes to the power of YouTube to change lives and businesses—it offers a place for every one of us to reach the whole world, no matter where we live or what our circumstances are. If you skin your YouTube knees, just get up, brush off, analyze what happened and why, then continue making great content you're passionate about, whether it's vlogging or quilting or building your business. Find your audience and speak to them. Create for them, listen to them, and never sacrifice your message for anyone or anything. When something isn't working, look at your data and make smart adjustments to give your content the best chance to be discovered and watched. This is the YouTube Formula. Use it, and watch the magic happen.

APPENDIX: FREE *YOUTUBE FORMULA* BONUS COMPANION COURSE

I couldn't write a book about YouTube without providing video resources. So I created a free companion course to help guide you through *The YouTube Formula*. It includes bonus video content, worksheets, and all the resources and links mentioned in this book. I highly recommend you sign up.

Visit the following link to get free access today:

www.ytformulabook.com/go

ABOUT THE AUTHOR

Derral Eves is one of the world's leading YouTube experts. He is CEO of Creatus, a video marketing and Strategy Company, and the founder of VidSummit, an annual event in Los Angeles for video creators and marketers. He has helped 24 YouTube channels go from zero to more than a million subscribers, and he has generated 54 billion views on YouTube.

Derral is founder, CEO and executive producer of *The Chosen*, the highest grossing crowdfunded movie project of all time. He was also executive producer on several viral video campaigns, including Squatty Potty's ice cream-pooping unicorn ad. Derral works with some of the biggest YouTube creators in the world, including MrBeast, who has tens of millions of subscribers and one of the fastest growing channels on YouTube.

Derral has been featured on *The Today Show*, *Good Morning America*, NBC, ABC, CBS, FOX, ESPN, FORBES, World Religion News, and more. He was featured on the Forbes list "20 Must Watch YouTube Channels That Will Change Your Business."

Derral's career passion and personal mission is to help individuals, brands, and businesses make a positive impact in the world, but his greatest passion is his family: his wife, Carolyn, and their five amazing children, Ellie, Logan, Kelton, Thatcher, and Bridger.

THIS IS NOT YOUR NORMAL ACKNOWLEDGEMENT

I want to acknowledge you, the reader. You are an important part of why I do what I do.

Let me tell you a quick story about our purpose:

When I was 12 years old, I went camping with my grandpa Jack. We were camping on top of a mountain by a reservoir overlooking the red rocks of Zion National Park in Utah. It was the most beautiful place to camp. Early in the morning, I was asleep in our tent when Grandpa Jack woke me before sunrise and asked me to go for a walk around the reservoir with him. The calm water looked like a sheet of glass at that early hour. It was so peaceful as the first light of day started to peak over the majestic mountains of Zion. Then a beam of light shot over the mountains, reflecting on the water, and it was the most beautiful, serene moment. At that moment, Grandpa Jack bent down and picked up a rock. With the rock in hand, he said, "Derral, you are like this rock. You are hard headed, strong, and a little rough around the edges. The water is the world . . ." and he threw the rock into the water, shattering the glassy surface and sending ripples across the reservoir. "What you do can impact so many people," Grandpa Jack said. "What impact will you have on the world? Will it be good, or will it be bad?" And he walked away, leaving me alone to contemplate the life lesson he had just taught. I had been camping with my grandpa many times in my life, and he had never done anything like this before. It felt so out of the ordinary that it left me a little dumbfounded. As I said, I was only 12 years old, so the power of the message didn't fully hit me at the time, but it continued to stick with me over the years. I've come back to it many times in my life. Thank you, Grandpa Jack, for the most powerful lesson that impacted me on every level. It continues to influence decisions I make in business and in my personal life.

Now, back to YOU. You are your own rock. Your content and your message will impact the world. My challenge to you is to use what I've taught you in this book to impact the world in a positive way. Entertain, inspire, motivate, educate, and elevate humanity with YouTube. Surround yourself with people who push you and motivate you and help you impact people's lives for good.

I want to acknowledge my wife, Carolyn, for her constant support, love, and patience with all of my craziness over the years. To my kids, Ellie, Logan, Kelton, Thatcher, and Bridger, for supporting a dad who doesn't have the normal nine-to-five job. Hopefully the perks of hanging out with their favorite YouTubers and traveling the world have been a good tradeoff.

To my parents for always facilitating my unique way of learning the way I needed to learn. You have been a constant source of support and strength for me.

To Melissa Young for controlling my ADHD and corralling my thoughts to write this book, and for being a constant supporter of every crazy project of mine. I also want to acknowledge her husband and kids for allowing her the space and time away at the whim of my crazy schedule. We had long hours and days away from our families, and we couldn't have finished the book without their support.

A big thank you to my business partners and my incredible team who helped me meet deadlines and modified meetings and tasks while I finished this book.

Last, I want to thank my third grade teacher for making me stand in a corner for a whole week. Even though you used it to try to humiliate me, what it did was ignite my pilot light and fuel my fire to be who I am. It taught me that I didn't have to conform to the standard to succeed and led me to where I am today.

INDEX